T0233781

SpringerBriefs in Mathematics

SpringerBriefs present concise summaries of cutting-edge research and practical applications across a wide spectrum of fields. Featuring compact volumes of 50 to 125 pages, the series covers a range of content from professional to academic. Briefs are characterized by fast, global electronic dissemination, standard publishing contracts, standardized manuscript preparation and formatting guidelines, and expedited production schedules.

Typical topics might include:

A timely report of state-of-the art techniques A bridge between new research results, as published in journal articles, and a contextual literature review A snapshot of a hot or emerging topic An in-depth case study A presentation of core concepts that students must understand in order to make independent contributions

SpringerBriefs in Mathematics showcases expositions in all areas of mathematics and applied mathematics. Manuscripts presenting new results or a single new result in a classical field, new field, or an emerging topic, applications, or bridges between new results and already published works, are encouraged. The series is intended for mathematicians and applied mathematicians. All works are peer-reviewed to meet the highest standards of scientific literature.

Titles from this series are indexed by Scopus, Web of Science, Mathematical Reviews, and zbMATH.

More information about this series at https://link.springer.com/bookseries/10030

http://www.sbmac.org.br/

Kaïs Ammari • Farhat Shel

Stability of Elastic Multi-Link Structures

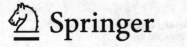 Springer

Kaïs Ammari
Department of Mathematics
University of Monastir
Monastir, Tunisia

Farhat Shel
Department of Mathematics
University of Monastir
Monastir, Tunisia

ISSN 2191-8198 ISSN 2191-8201 (electronic)
SpringerBriefs in Mathematics
ISBN 978-3-030-86350-0 ISBN 978-3-030-86351-7 (eBook)
https://doi.org/10.1007/978-3-030-86351-7

This Springer imprint is published by the registered company Springer Nature Switzerland AG
The registered company address is: Gewerbestrasse 11, 6330 Cham, Switzerland

Preface

In this book, we investigate the asymptotic behavior of some PDEs on networks. The structures we consider consist of finitely interconnected flexible elements such as strings and beams or combinations thereof, distributed along a planar network.

Such a study is motivated by the need for engineers to eliminate vibrations in some dynamical structures consisting of elastic bodies, coupled in the form of chain or graph such as pipelines, bridges, and some cable networks. There are other complicated examples in the automotive industry, aircraft and space vehicles, containing rather than strings and beams, plates and shells. These multi-body structures are often complicated; moreover, the mathematical models describing their evolution are quite complex. For the sake of simplicity, we consider only $1 - d$ networks.

We use semigroup approach to investigate well-posedness and stability. We write the dynamical system as an abstract linear evolution equation on an appropriate Hilbert space H:

$$\frac{du}{dt}(t) = \mathcal{A}u(t), \quad u(0) = u^0, \tag{*}$$

where $\mathcal{A}$ is an unbounded operator on H (state space); the Hille–Yosida theorem will be a principal tool for proving existence, uniqueness, and regularity of the solution of (*). Moreover, in some way, the exponential (or polynomial) stability of the (physical) system is viewed as the exponential (or polynomial) stability of the corresponding semigroup, then our main tool for studying exponential decay is the frequency domain characterization (Theorem 1.24 or precisely Theorem 1.25 in this book) due to Gearhart [28], Pruss [80], and [42]. For proving polynomial stability, we use an analogous theorem (Theorem 1.26 in this book), due to Borichev and Tomilov [14].

In order that this book be self-contained, we start with an introductory chapter (Chap. 1), in which we first present some elementary notions on graph theory, in relation to the study of partial differential equations on networks. Second, we recall some basic definitions and theorems about semigroup theory, and we give a brief reminder about Sobolev spaces and interpolation spaces.

The rest will be divided into two parts: the first, Part I, includes the first two chapters and deals with networks of elastic materials, some of which are thermoelastic.

In Chap. 2, we consider a network of strings under two cases: in the first, two elastic edges cannot be adjacent, while in the second, the network is a tree of elastic strings, the leaves of which thermoelastic edges are added. We prove the exponential stability of the C_0-semigroup associated with the initial boundary value problem.

In Chap. 3, we consider a network of beams, and we assume that the graph has at least an external node, that every maximal subgraph of elastic edges is a tree, the leaves of which thermoelastic edges are attached, and that every maximal subgraph of thermoelastic edges is not a circuit. It is proved that the corresponding semigroup associated to some initial and boundary conditions is exponentially stable.

The second part of this book includes Chaps. 4–6.

In Chap. 4, we consider a tree of elastic strings and beams, where only external controls are applied at leaves (boundary feedback), we prove that the whole system is exponentially stable if there is no beam following a string (from the root to the leaves) and polynomially stable if at least a beam follows a string.

Chapter 5 is devoted to studying the stability of a model of fluid propagation in a $1 - d$ tree-shaped network, under some feedbacks applied at exterior nodes, with the presence of point mass at inner nodes. It is proved that under some (algebraic) conditions on lengths of edges not attached to leaves, the corresponding semigroup is exponentially stable. In the last section, we discuss the case where the graph contains a circuit and the case where at least a control is omitted.

In Chap. 6, we consider a network of elastic strings with local Kelvin–Voigt damping on some edges, and we prove, under some condition on the regularity of the damping coefficient function, some results of polynomial and exponential stability of the associated semigroup.

Monastir, Tunisia   Kaïs Ammari

Monastir, Tunisia   Farhat Shel
November 2020

Contents

Chapter 1
Preliminaries

1 Introduction

In this chapter, we summarize some definitions and results used in this book or serve to better understand the content in such a way that this book will be self-contained.

We first introduce some definitions and notations on graphs or networks, as introduced in [1, 34, 68], or [8] (see also [18, 98]).

Second, we concentrate on some results about spectrum, resolvent set, and resolvents of (unbounded) linear operators on Hilbert spaces. We give the basics about strongly continuous semigroups of operators on Hilbert spaces. Then, we introduce the Hille–Yosida theorem giving a complete characterization of operators that generate strongly continuous semigroup and the well-known Lumer–Phillips theorem (for contraction semigroups). Next we give a positive answer about the well-posedness of an abstract Cauchy problem

$$\begin{cases} \dfrac{du}{dt}(t) = Au(t) \text{ for } t > 0, \\ u(0) = x, \end{cases} \tag{1.1}$$

where $A : D(A) \subset X \to X$ is a linear operator generating a strongly continuous semigroup $(T(t))_{t \geq 0}$ on a Hilbert space X, and $x \in X$ the initial value.

For all this, we can cite [23, 93], and the references therein, in particular [15, 19, 20, 30, 41, 79, 103].

Afterward, we focus on the asymptotic stability of the semigroup $(T(t))_{t \geq 0}$. We present different concepts of stability (strong, exponential, and polynomial), and we give a characterization of each one in terms of its generator and its resolvent. For more details, see [13, 14, 23, 28, 42, 101].

We end this chapter with a brief reminder on Sobolev spaces in one dimension, in particular, Sobolev embedding theorems and Gagliardo–Nirenberg and Poincaré inequalities.

© The Author(s), under exclusive license to Springer Nature Switzerland AG 2022
K. Ammari, F. Shel, *Stability of Elastic Multi-Link Structures*, SpringerBriefs in
Mathematics, https://doi.org/10.1007/978-3-030-86351-7_1

Throughout this chapter, X is a real or complex Hilbert space with the inner product $\langle .,. \rangle$ and the induced norm $\|.\|$. We also use the notation $\|.\|$ to denote the (induced) norm in $\mathcal{L}(X) := \mathcal{L}(X, X)$, the space of bounded linear operators in X.

2 Terminology of Networks

We first introduce some notations needed to formulate the problems under consideration (see, for example, [8, 18, 34] for notations used in this book). Let $\mathcal{G}$ be a planar metric connected graph embedded in $\mathbb{R}^m$, $m \in \mathbb{N}^* := \mathbb{N} \setminus \{0\}$, with N edges $e_1, \ldots, e_N$, $N \geq 1$ and p vertices $a_1, \ldots, a_p$, $p \geq 2$. When the graph is a tree, it will be denoted by $\mathcal{T}$ instead of $\mathcal{G}$, and note that in this case $p = N + 1$.

By degree (or valency) of a vertex of $\mathcal{G}$ we mean the number of edges incident at the vertex. If the degree is equal to one, the vertex is called exterior; otherwise, it is said to be interior. Denote the set of boundary vertices by $\mathcal{V}_{ext}$ and the set of vertices with degree more than one by $\mathcal{V}_{int}$. Moreover, we denote by I_{int} and I_{ext}, respectively, the sets of indices of interior and exterior vertices, and then $I := \{1, \ldots, p\} = I_{int} \cup I_{ext}$ is the set of indices of all vertices.

We set $J := \{1, \ldots, N\}$, and for $k \in I$, we will denote by J_k the set of indices of edges adjacent to the vertex a_k. If $k \in I_{ext}$, then the unique element of J_k will be denoted by j_k. Finally, the notation $s(i, k) = s(k, i)$ denotes the index of the edge connecting a_i and a_k.

The length of the edge e_j is denoted by ℓ_j. Then, e_j may be parametrized by its arc length by means of the functions $\pi_j : [0, \ell_j] \longrightarrow e_j$, $x \longmapsto \pi_j(x)$. Sometimes, we identify e_j with the interval $[0, \ell_j]$.

For a function $u : \mathcal{G} \longrightarrow \mathbb{C}$, we set $u_j = u \circ \pi_j$ its restriction to the edge e_j. For simplicity, we will write $u = (u_1, \ldots, u_N)$, and we will use the abbreviation $u_j(x) = u(\pi_j(x))$ for any $x \in (0, \ell_j)$. For a node a in the graph $\mathcal{G}$, which is an end of an edge e_j, we will often write $u_j(a)$ instead of $u_j(\pi_j^{-1}(a))$ and $u_{j,x}(a)$ instead of $u_{j,x}(\pi_j^{-1}(a))$. Moreover, $u_j(a)$ will be sometimes written simply as $u(a)$ if u is continuous at a, etc.

The orientation of the graph G is given by the incidence matrix $D = (d_{kj})_{p \times N}$, defined by,

$$d_{kj} = \begin{cases} 1 & \text{if } \pi_j(\ell_j) = a_k, \\ -1 & \text{if } \pi_j(0) = a_k, \\ 0 & \text{otherwise.} \end{cases}$$

Define the adjacency matrix $E = (e_{ik})_{p \times p}$ of $\mathcal{G}$ by

$$e_{ik} = \begin{cases} 1 & \text{if } a_i \text{ and } a_k \text{ are adjacent,} \\ 0 & \text{otherwise.} \end{cases}$$

The Hadamard product of two matrices $A = (a_{jk})$ and $B = (b_{jk})$ of the same size is defined as

$$A * B = (a_{jk}b_{jk}),$$

and for any function $Q : \mathbb{R} \longrightarrow \mathbb{R}$, we define the matrix $Q(A) = (q_{ik})_{p \times p}$ by

$$q_{ik} = \begin{cases} Q(a_{ik}) & \text{if } e_{ik} = 1, \\ 0 & \text{otherwise.} \end{cases}$$

In particular, we write $A^{(r)} = Q(A)$ if $Q(x) = x^r$. Furthermore, denote by L the matrix $L = (\ell_{ik})_{p \times p}$, where

$$\ell_{ik} = \begin{cases} \ell_{s(i,k)} & \text{if } e_{ik} = 1, \\ 0 & \text{otherwise,} \end{cases}$$

where $s(i, k) = s(k, i)$ is the index of the edge connecting a_i and a_k.

3 Spectrum and Resolvents of an Operator

Definition 1.1 A linear operator $A : D(A) \subset X \mapsto X$ is called *closed* if its graph, defined by $G(A) = \{(x, Ax) \mid x \in D(A)\}$, is closed in $X \times X$.

Clearly, A is closed if and only if for any sequence x_n in $D(A)$ such that $x_n \to x$ in X and $Ax_n \to y$ in X, we have $x \in D(A)$ and $y = Ax$.

Note that if $A : D(A) \subset X \mapsto X$ is closed and $B \in \mathcal{L}(X)$, then also $A + B$ is closed (the domain of $A + B$ is $D(A)$).

Definition 1.2 Let $A : D(A) \subset X \to X$ be a linear operator.

- The *resolvent set* of A, denoted by $\rho(A)$, is the set of those points $\lambda \in \mathbb{C}$, for which the operator $\lambda I - A : D(A) \to X$ is invertible and $(\lambda I - A)^{-1} \in \mathcal{L}(X)$. The *spectrum* of A, denoted by $\sigma(A)$, is the complement of $\rho(A)$ in $\mathbb{C}$. For $\lambda \in \rho(A)$, $R(\lambda, A) := (\lambda I - A)^{-1}$ is called a *resolvent* of A.
- $\lambda \in \mathbb{C}$ is called an *eigenvalue* of A if the operator $\lambda I - A : D(A) \to X$ is not injective. In this case, a $z_\lambda \in D(A)$, $z_\lambda \neq 0$, satisfying $Az_\lambda = \lambda z_\lambda$ is called an *eigenvector* of A corresponding to λ. The set of all the eigenvalues of A is called the point spectrum of A, and it is denoted by $\sigma_p(A)$.

Note that if $\rho(A)$ is not empty, then A is closed. Indeed, let $\lambda \in \rho(A)$, and let x_n be a sequence in $D(A)$ such that $x_n \to x$ in X and $(A - \lambda I)x_n \to y$ in X. Since $(A - \lambda I)^{-1}$ is bounded, we have that $x_n = (A - \lambda I)^{-1}(A - \lambda I)x_n$ converges to $(A - \lambda I)^{-1}y$, which implies that $(A - \lambda I)^{-1}y = x$, and hence $x \in D(A)$ and $(A - \lambda I)x = y$.

Note also that it can be proved that $\rho(A)$ is open and then $\sigma(A)$ is closed.

In the following proposition, we show that for every $\lambda \in \rho(A)$ there is a canonical relation, called the spectral mapping theorem, between the unbounded operator A and the spectrum of the bounded operator $R(\lambda, A) = (\lambda I - A)^{-1}$.

Proposition 1.3 *Let* $A : D(A) \subset X \to X$ *be a linear operator with non-empty resolvent set* $\rho(A)$ *(in particular, A is closed). Then,*

(i) $\sigma((\lambda I - A)^{-1}) \setminus \{0\} = (\lambda - \sigma(A))^{-1} := \left\{ \frac{1}{\lambda - \mu}, \ \mu \in \sigma(A) \right\}.$

(ii) *Analogous statements hold for the point spectrum of A and $(\lambda I - A)^{-1}$.*

An important case is when the resolvent $(\lambda I - A)^{-1}$ is compact, for some $\lambda \in \rho(A)$ (we say that A has *compact resolvent*), we then have the following theorem [45]:

Theorem 1.4 *Let* $A : D(A) \subset X \to X$ *be an operator such that* $(\lambda I - A)^{-1}$ *is compact, for some* $\lambda \in \rho(A)$. *Then,* $\sigma(A)$ *is discrete and formed only of eigenvalues of finite multiplicity; in particular, we have* $\sigma(A) = \sigma_p(A)$. *Moreover,* $(sI - A)^{-1}$ *is compact for every* $s \in \rho(A)$.

For concrete operators, the following characterization is quite useful.

Proposition 1.5 *Let* $A : D(A) \subset X \to X$ *be an operator with* $\rho(A) \neq \emptyset$ *and take* $X_1 := (D(A), \|.\|_A)$. *Then, the following assertions are equivalent:*

(a) *The operator A has compact resolvent.*
(b) *The canonical injection* $i : X_1 \hookrightarrow X$ *is compact.*

4   Semigroups

Definition 1.6 A family $(T(t))_{t \geq 0}$ of bounded linear operators in the Hilbert space X is called strongly continuous semigroup or C_0-semigroup (or just semigroup) if the following conditions are fulfilled:

(a) $T(0) = I$, and we have $T(t + s) = T(t)T(s)$ for all $t, s \geq 0$.
(b) For each $x \in X$, the orbit, defined as the map

$$T(\cdot)x : \mathbb{R}_+ \to X, \quad t \mapsto T(t)x,$$

is continuous.

Property (a) is called the *semigroup law* and (b) is the *strong continuity*, and it can be replaced by

$$\lim_{t \to 0^+} T(t)x = x.$$

We said that $(T(t))_{t \geq 0}$ is *uniformly continuous* if the mapping $t \mapsto T(t)$ is continuous for the operator norm. It is proved that $(T(t))_{t \geq 0}$ is uniformly continuous if and only if there exists a bounded operator A on X such that

$$\forall t \geq 0, \quad T(t) = e^{tA} = \sum_{k=0}^{\infty} \frac{(tA)^k}{k!}.$$

The operator A is given by

$$A = \lim_{t \to 0^+} \frac{1}{t}(T(t) - I).$$

For a semigroup $(T(t))_{t \geq 0}$, we define an operator A by

$$D(A) = \left\{ x \in X \mid \text{the limit } \lim_{t \to 0^+} \frac{1}{t}(T(t)x - x) \text{ exists in } X \right\},$$

and for $x \in D(A)$,

$$Ax = \lim_{t \to 0^+} \frac{1}{t}(T(t)x - x).$$

Then, A is called the *infinitesimal generator* (or shortly, the *generator*) of the semigroup $(T(t))_{t \geq 0}$. We also say that A generates $(T(t))_{t \geq 0}$, and sometimes we denote $T(t)$ by e^{At}.

Some elementary properties are collected in the following proposition:

Proposition 1.7 *Let A be the generator of a C_0-semigroup $(T(t))_{t \geq 0}$ on X. The following properties hold:*

(i) *The map $A : D(A) \subset X \to X$ is effectively a linear operator.*
(ii) *If $x \in D(A)$, then $T(t)x \in D(A)$ and*

$$\frac{d}{dt}T(t)x = T(t)Ax = AT(t)x, \quad \text{for all } t \geq 0.$$

(iii) *For every $t \geq 0$, one has*

$$T(t)x - x = A \int_0^t T(t)x \, ds \quad \text{if } x \in X,$$

$$= \int_0^t T(t)Ax \, ds \quad \text{if } x \in D(A).$$

Theorem 1.8 *The generator of a C_0-semigroup is closed and densely defined linear operator that determines the semigroup uniquely.*

We observe that every C_0-semigroup is exponentially bounded.

Lemma 1.9 *Let $(T(t))_{t\geq 0}$ be a C_0-semigroup on X. There exist $M \geq 1$ and $\omega \in \mathbb{R}$ such that*

$$\forall\, t \geq 0, \quad \|T(t)\| \leq Me^{\omega t}.$$

Moreover, a semigroup $(T(t))_{t\geqslant 0}$ is called *bounded* if we can take $\omega = 0$ and contractive (or *semigroup of contractions*) if $\omega = 0$ and $M = 1$ is possible, that is, $\|T(t)\| \leq 1$ for all $t \geq 0$.

Definition 1.10 Let $(T(t))_{t\geq 0}$, a C_0-semigroup with generator A. Then,

$$\omega_0(T) := \omega_0(A) := \inf\big\{\omega \in \mathbb{R} \mid \exists M_\omega \geq 1 : \|T(t)\| \leq M_\omega e^{\omega t} \text{ for all } t \geq 0\big\}$$
$$= \inf\big\{\omega \in \mathbb{R} \mid t \mapsto e^{-\omega t}\|T(t)\| \text{ is bounded on } \mathbb{R}_+\big\}$$

is called the *growth bound* of $(T(t))_{t\geq 0}$.

Note that $\omega_0(T) \in [-\infty, +\infty)$.

Proposition 1.11 *Let $(T(t))_{t\geq 0}$ be a C_0-semigroup with generator A. For every $w > \omega_0(T)$, there exists $M_\omega \in [1, +\infty)$ such that*

$$\|T(t)\| \leq M_\omega e^{\omega t} \quad \text{for all } t \geq 0.$$

5 Hille–Yosida Generation Theorems

5.1 *Generation of Semigroups*

We want to characterize those linear operators that are the generators of some C_0-semigroup.

First, for every $\omega \in \mathbb{R}$, we define the right half-plane $\mathbb{C}_\omega := \{\lambda \in \mathbb{C} \text{ with } \mathrm{Re}\,\lambda > \omega\}$, in particular, $\mathbb{C}_0 := \{\lambda \in \mathbb{C} \text{ with } \mathrm{Re}\,\lambda > 0\}$.

Theorem 1.12 (Hille–Yosida Theorem, Contraction Case, Hille, Yosida 1948 [40, 102]) *Let A be a linear operator on X. The following assertions are equivalent:*

(i) The operator A is the generator of a C_0-semigroup of contractions.
(ii) A is closed, $D(A)$ is dense in X, $(0, \infty) \subset \rho(A)$, and

$$\|(\lambda I - A)^{-1}\| \leq \frac{1}{\lambda} \quad \text{for } \lambda > 0.$$

(iii) A is closed, $D(A)$ is dense in X, $\mathbb{C}_0 \subset \rho(A)$, and

$$\|(\lambda I - A)^{-1}\| \leq \frac{1}{Re\lambda} \quad for\ \lambda \in \mathbb{C}_0.$$

The above result can be generalized to any $\mathcal{C}_0$-semigroup as follows:

Theorem 1.13 (Hille–Yosida Theorem) *Let A be a linear operator on X. The following assertions are equivalent:*

(i) A is the generator of a $\mathcal{C}_0$-semigroup $(T(t))_{t\geq 0}$ with $\|T(t)\| \leq Me^{\omega t}$, for $t \geq 0$, for some $\omega \in \mathbb{R}$ and some $M \geq 1$.
(ii) A is closed, $D(A)$ is dense in X, $(\omega, \infty) \subset \rho(A)$, and

$$\|(\lambda I - A)^{-1}\| \leq \frac{M}{\lambda - \omega} \quad for\ \lambda > \omega.$$

(iii) A is closed, $D(A)$ is dense in X, $\mathbb{C}_\omega \subset \rho(A)$, and

$$\|(\lambda I - A)^{-1}\| \leq \frac{M}{Re\lambda - \omega} \quad for\ \lambda \in \mathbb{C}_\omega.$$

5.2 Dissipative Operators and Contraction Semigroups

Definition 1.14 An operator $A : D(A) \subset X \to X$ is said to be dissipative if for any $x \in D(A)$,

$$Re(\langle Ax, x \rangle) \leq 0.$$

An operator A is said to be m-dissipative (or maximal dissipative) if it is dissipative and there is a $\lambda_0 > 0$ such that $Ran(\lambda_0 I - A) = X$.

For more details about the following theorem, see [79].

Theorem 1.15 (Lumer-Phillips) *For any linear operator $A : D(A) \subset X \to X$, the following statements are equivalent:*

(i) A is the generator of a $\mathcal{C}_0$-semigroup of contractions.
(ii) A is m-dissipative.

As a corollary of the above theorem (using that $\rho(A)$, the resolvent set of A, is open in $\mathbb{C}$), the following result will be frequently used in this book.

Theorem 1.16 *Let $A : D(A) \subset X \to X$ be a linear operator on X. If A is dissipative and $0 \in \rho(A)$, then A is the generator of a $\mathcal{C}_0$-semigroup of contractions.*

6 Abstract Cauchy Problems

We consider the abstract Cauchy problem

$$
\begin{cases}
\dfrac{du(t)}{dt} = Au(t) \text{ for } t > 0, \\
u(0) = x,
\end{cases}
\tag{1.2}
$$

where the independent variable t represents time, $u(.)$ is a function with values in the Hilbert space X, $A : D(A) \subset X \to X$ a linear operator, and $x \in X$ the initial value.

A continuous function $u : \mathbb{R}_+ \to X$ is called a (strong) *solution* of (1.2) if u is continuously differentiable with respect to X, $u(t) \in D(A)$ for all $t \geq 0$, and (1.2) holds.

A continuous function $u : \mathbb{R}_+ \to X$ is called a (mild) *solution* of (1.2) if $\int_0^t u(s)ds \in D(A)$ for all $t \geq 0$ and $u(t) = A \int_0^t u(s)ds + x$.

The following theorem ensures the existence and uniqueness of a solution of the initial value problem (1.2).

Theorem 1.17 *If the operator A generates a C_0-semigroup, then for any initial datum $x \in X$, there exists a unique (mild) solution u of the abstract Cauchy problem (1.2). u satisfies*

$$
u \in \mathcal{C}([0, \infty), X).
$$

Moreover, for $x \in D(A)$, there exists a unique (strong) solution u of the abstract Cauchy problem (1.2). u satisfies

$$
u \in \mathcal{C}([0, \infty), D(A)) \cap \mathcal{C}^1([0, \infty), X).
$$

In the two cases, $u(t) = T(t)x$, where $(T(t))_{t \geq 0}$ is the C_0-semigroup generated by A.

7 Stability

We assume that the operator A in (1.2) generates a C_0-semigroup $(T(t))_{t \geq 0}$. Then, in general, we define the energy E of the solution u of the abstract Cauchy problem (1.2) as

$$
E(t) = \frac{1}{2} \|u(t)\|^2 = \frac{1}{2} \|T(t)x\|^2,
$$

and in fact, it is the case in all the dynamic systems considered in this book.

In studying a dynamic system, one important thing is to study the asymptotic behavior of its semigroup $T(t)$ or, equivalently, its energy (for large $t > 0$). In this book, we focus on the stability, by this we mean that the energy (or the semigroup) converges to zero as $t \to \infty$.

We distinguish different concepts of convergences.

Definition 1.18 We assume that A generates a C_0-semigroup $(T(t))_{t \geq 0}$ on X.

(a) We say that $(T(t))_{t \geq 0}$ is strongly (or asymptotically) stable if

$$\lim_{t \to \infty} \|T(t)x\| = 0 \quad \text{for all } x \in X. \tag{1.3}$$

(b) We say that $(T(t))_{t \geq 0}$ is exponentially (or uniformly) stable if there exist constants $\alpha > 0$ and $M \geq 1$ such that

$$\|T(t)\| \leq Me^{-\alpha t} \quad \text{for all } t \geq 0. \tag{1.4}$$

Remark 1.19 Property (b) is equivalent to

(b′)

$$\lim_{t \to \infty} \|T(t)\| = 0$$

.

Indeed, the implication (b) $\Rightarrow$ (b′) is trivial. Conversely, suppose that (b′) holds, there exists $t_0 > 0$ such that $q := \|T(t_0)\| < 1$. We set $M = \sup_{0 \leq s \leq t_0} \|T(s)\|$, which exists, since $t \mapsto \|T(t)\|$ is continuous. If we decompose $t = kt_0 + s \in \mathbb{R}_+$ with $k \in \mathbb{N}$ and $s \in [0, t_0)$, we obtain

$$\|T(t)\| \leq \|T(s)\| \cdot \|T(kt_0)\| \leq M\|T(t_0)^k\|$$

$$\leq Mq^k = Me^{k\ln(q)} \leq \frac{M}{q}e^{-\alpha}t,$$

where $\alpha := -\frac{\ln(q)}{t_0}$. Thus, (b) holds.

Note also that exponential stability implies strong stability.

In (a), we will also say that the system (1.2) is strongly stable, and it also means that

$$\lim_{t \to \infty} E(t) = 0 \quad \text{for all } x \in X,$$

where $E(t) := \frac{1}{2}\|T(t)x\|^2$.

In (b), condition (1.4) can be rewritten as

$$E(t) \leq CE(0)e^{-\delta t} \quad \text{for all } t > 0,$$

where δ and C are positive constants (not depending on the initial condition). We will also say that the system (1.2) (or the solution or the energy of the system (1.2)) is exponentially stable.

In the literature, there is another concept of stability "between" the previous two, called polynomial stability, which is often written for the energy of the following form:

$$E(t) \leq \frac{C}{t^\alpha} \|x\|^2_{D(A)}, \quad \text{for all } x \in D(A) \text{ and for all } t > 0,$$

where C and α are positive constants and where $\|x\|_{D(A)} = \|x\| + \|Ax\|$. Recall here that $E(t)$ is the energy of the solution u satisfying the initial condition $u(0) = x$. This translates in a more precise way as follows:

Definition 1.20 We assume that A generates a C_0-semigroup $(T(t))_{t \geq 0}$ on X. We say that $(T(t))_{t \geq 0}$ is polynomially stable if there exist two positive constants α and M such that, for every $x \in D(A)$,

$$\|T(t)x\| \leq \frac{M}{t^\alpha} \|x\|_{D(A)} \quad \text{for all } t > 0. \tag{1.5}$$

For more details on polynomial stability, one can see, for example, [14, 55], and the references therein. In this book, we choose to characterize the stability of a semigroup in terms of its generator and its resolvent.

7.1   Strong Stability

First, we give some sufficient conditions of strong stability of C_0-semigroups.

Theorem 1.21 (Arendt and Batty [13]) *Let $(T(t))_{t \geq 0}$ be a C_0-semigroup of contractions, with generator A on the Hilbert space X. Denote by $\sigma(A)$ the spectrum of A. If $\sigma(A) \cap i\mathbb{R}$ is at most countable and no eigenvalue of A lies on the imaginary axis, then $(T(t))_{t \geq 0}$ is strongly stable.*

Corollary 1.22 *Let $(T(t))_{t \geq 0}$ be a C_0-semigroup of contractions, with generator A on the Hilbert space X. If $i\mathbb{R} \subset \rho(A)$, then $(T(t))_{t \geq 0}$ is strongly stable.*

7.2   Exponential Stability

First, it is clear that a C_0-semigroup, generated by an operator A, is exponentially stable if and only if

$$\omega_0(A) := \omega_0(T) < 0.$$

Next, as mentioned below, we want to characterize the exponential stability of the semigroup in terms of its generator. For this purpose, define the spectral bound of an operator A as follows:

$$s(A) := \sup\{Re\lambda, \quad \lambda \in \sigma(A)\}.$$

We know that in the case of finite dimension, the C_0-semigroup e^{tA} is exponentially stable if and only if

$$s(A) < 0. \tag{1.6}$$

This condition, although necessary, is not sufficient in general to have exponential stability in infinite dimension, nevertheless it is necessary and sufficient for a class of semigroups (see [23]).

Proposition 1.23 *For a C_0-semigroup with generator A, one has*

$$-\infty \leq s(A) \leq \omega_0(A) < +\infty.$$

In particular, if $s(A) = 0$, then the C_0-semigroup with generator A is not exponentially stable.

Now we introduce the main result used for proving exponential stability of C_0-semigroups in this book [28, 42, 80].

Theorem 1.24 (Gearhart [28], Prüss [80], and Huang [42]) *Let $(T(t))_{t\geq 0}$ be a C_0-semigroup on a Hilbert space X, with generator A. Then, $(T(t))_{t\geq 0}$ is exponentially stable if and only if*

$$\{\lambda \in \mathbb{C}, \quad Re\lambda > 0\} \subset \rho(A), \tag{1.7}$$

and

$$\sup_{Re\lambda>0} \|(\lambda I - A)^{-1}\|_{\mathcal{L}(X)} < \infty. \tag{1.8}$$

Note that condition (1.7) can be replaced by $s(A) := \sup\{Re\lambda, \quad \lambda \in \sigma(A)\} < 0$. Moreover, the inequality $Re\lambda > 0$ in (1.8) can be large.

The following invariant of the result is due to Gearhart [28].

Theorem 1.25 *Let $(T(t))_{t\geq 0}$ be a C_0-semigroup of contractions on a Hilbert space X, with generator A. Then, $(T(t))_{t\geq 0}$ is exponentially stable if and only if*

$$i\mathbb{R} \subset \rho(A) \tag{1.9}$$

and

$$\limsup_{\beta \in \mathbb{R}, |\beta| \to +\infty} \left\| (i\beta I - A)^{-1} \right\|_{\mathcal{L}(X)} < \infty. \tag{1.10}$$

Note that condition (1.10) can be replaced by $\sup_{w \in \mathbb{R}} \|(iwI - A)^{-1}\|_{\mathcal{L}(X)} < \infty$.

7.3   *Polynomial Stability*

The following characterization of polynomial stability of a C_0-semigroup of contractions is due to Borichev and Tomilov [14].

Theorem 1.26 *A C_0-semigroup of contraction $(e^{tA})_{t\geq0}$ on the Hilbert space X such that $i\mathbb{R} \subset \rho(A)$ satisfies, for every $x \in D(A)$,*

$$\|e^{tA}x\| \leq \frac{M}{t^\alpha}\|x\|_{D(A)} \quad \textit{for all } t > 0, \tag{1.11}$$

for some constant $M > 0$ and for $\alpha > 0$ if and only if

$$\limsup_{\beta\in\mathbb{R},|\beta|\to+\infty} \frac{1}{|\beta|^{\frac{1}{\alpha}}} \left\|(i\beta I - A)^{-1}\right\|_{\mathcal{L}(X)} < \infty. \tag{1.12}$$

8   Sobolev Spaces in One Dimension

In this section, we collect some basic results on function spaces, which will be used frequently in this book. For more details, we refer the reader to [2, 15, 52], and the first Chapter in [57]. See also [25, 31, 74] and in particular the Appendix in [93] and the references therein.

8.1   *Definition and First Properties*

First, we briefly recall the definition and main results of the Sobolev spaces $H^m(\Omega)$, where $\Omega =]a, b[$ such that $a, b \in \mathbb{R}$ and $a < b$.

We denote by $C_0^\infty(\Omega)$ or $\mathcal{D}(\Omega)$ the set of all φ in $C^\infty(\Omega)$, which have compact support contained in Ω.

We define the space

$$H^1(\Omega) := \left\{u \in L^2(\Omega) \mid \partial_x u \in L^2(\Omega)\right\}.$$

In other words, a function $u \in L^2(\Omega)$ is in $H^1(\Omega)$ if and only if there exists $v \in L^2(\Omega)$ such that

$$\int_\Omega u\partial_x\bar{\varphi}dx = -\int_\Omega v\bar{\varphi}dx$$

for every φ in $C_0^\infty(\Omega)$. The function v is denoted by $\partial_x u$.

With the inner product defined by

$$\langle u, v \rangle = \int_\Omega u \bar{v} dx + \int_\Omega \partial_x u \partial_x \bar{v} dx,$$

$H^1(\Omega)$ is a Hilbert space.

In the same way, we define the Sobolev spaces $H^m(\Omega)$, where $m \in \mathbb{N}$ by

$$H^m(\Omega) := \left\{ u \in L^2(\Omega) \mid \partial_x^\alpha u \in L^2(\Omega), \ \alpha \in \mathbb{N}, \ \alpha \leq m \right\}.$$

From the above definition, it clearly follows that $H^0(\Omega) = L^2(\Omega)$. Equipped with the inner product:

$$\langle u, v \rangle = \sum_{\alpha=0}^m \int_\Omega \partial_x^\alpha u \partial_x^\alpha \bar{v} dx,$$

$H^m(\Omega)$ is a Hilbert space.

Theorem 1.27 (Density Theorem) *For $m \geq 1$, $C^m(\bar{\Omega})$ is dense in $H^m(\Omega)$.*

Theorem 1.28 *For $m > \frac{1}{2} + k$, $H^m(\Omega) \subset C^k(\bar{\Omega})$.*

Remark 1.29 Any function $u \in H^1(\Omega)$ admits a unique representative continuous on $\bar{\Omega}$, still denoted by u, and we have, for every $x, y \in \bar{\Omega}$,

$$u(y) = u(x) + \int_x^y \partial_x u(t) dt.$$

Proposition 1.30 (Integration by Parts) *If $u, v \in H^1(\Omega)$, then $uv \in H^1(\Omega)$ and $\partial_x(uv) = \partial_x u \cdot v + u \partial_x v$. In particular, for all $[x, y] \subset [a, b]$,*

$$\int_{[x,y]} (\partial_x u \cdot v + u \partial_x v) = (uv)(y) - (uv)(x).$$

8.2 Compact Embeddings, $H_0^1(\Omega)$ Space

Theorem 1.31 *For $0 \leq m_1 < m_2$, $H^{m_2}(\Omega) \subset H^{m_1}(\Omega)$ with compact embedding.*

We define the space $H_0^1(\Omega)$ as the closure of $C_0^\infty(\Omega)$ in $H^1(\Omega)$. We have

$$H_0^1(\Omega) := \left\{ u \in H^1(\Omega) \mid u(a) = u(b) = 0 \right\}.$$

Theorem 1.32 *The embedding operator $H_0^1(\Omega) \hookrightarrow L^2(\Omega)$ is compact.*

8.3   Some Useful Inequalities

The following inequalities are frequently used in this book.

The firsts two inequalities can be found in a more general case in [57] (see also [26, 76]).

Theorem 1.33 (Some Gagliardo–Nerenberg Inequalities)

(1) There are two positive constants C_1 and C_2 such that for any w in $H^1(\Omega)$,

$$\|w\|_\infty \leq C_1 \|\partial_x w\|^{1/2} \|w\|^{1/2} + C_2 \|w\|. \tag{1.13}$$

(2) There are two positive constants C_3 and C_4 such that for any w in $H^2(\Omega)$,

$$\|\partial_x w\| \leq C_3 \left\|\partial_x^2 w\right\|^{1/2} \|w\|^{1/2} + C_4 \|w\|. \tag{1.14}$$

The following inequality is known as the Poincaré inequality:

Theorem 1.34 *There exists a positive constant C, depending only on Ω such that*

$$\|u\| \leq C \|\partial_x u\|, \quad \forall u \in H_0^1(\Omega)$$

(the constant $C = |b - a|$ is suitable).

9   Comments

Stability of PDEs (in particular, stability of beam and string systems) can be found in many books such as Z.-H. Liu et al. [58], J. Oostveen [78], M. Krstic et al. [47], T. Meurer [70], A. Zuyev [108], M. Gugat [32], G. Sklyar et al. [91], and [18].

For interested people, there are some recent developments in the field of differential equations on networks, such as [64, 71, 105], on some fractional boundary value problems on graphs.

Chapter 2
Exponential Stability of a Network of Elastic and Thermoelastic Materials

The wave equation on an elastic body is conservative; to make the system stable, several authors have introduced different types of dissipative mechanisms, for example, a frictional damping [38] or frictional boundary conditions [46, 77]. For the stabilization of a network governed by wave equations, we refer to [4, 10], where the authors considered a star-shaped and tree-shaped networks of elastic strings, and they proved that when a feedback is applied on particular nodes, the system will be polynomially stable but not exponentially stable. We can see, in [75], that the authors considered a network with delay term in the nodal feedbacks. In [12], the authors studied, in particular, the stabilization of a chain of beams and strings; see also [33–35, 94].

Another type of stabilization of an elastic material is to add thermoelastic materials to it. In [61, 62, 81], the authors proved that the system is then exponentially stable; see also [84] where the authors considered the case of beams and proved that the whole system is also exponentially stable. We want to know if this result holds true for a network of elastic and thermoelastic materials. To our knowledge, the asymptotic behavior of such a system has not been studied yet. In this chapter, we consider particular cases of such network that can be partially generalized. In the first case, we suppose that two elastic edges cannot be adjacent (Fig. 2.1). In the second one, we consider a tree of elastic materials, the leaves of which thermoelastic materials are added as follows: the thermoelastic body is related to only one leaf by an end, and the second is free or connects two leaves, with the condition that each leaf is connected to only one thermoelastic body (Fig. 2.2). With the continuity condition for the displacement and the Neumann condition for the temperature at the internal nodes, we prove that the thermal effect is strong enough to stabilize the system. We will use a frequency method as described in the introduction.

We consider a network of elastic and thermoelastic bodies, with N edges $e_1, \ldots, e_N$ and p vertices $a_1, \ldots, a_p$, and that coincides with the graph $\mathcal{G}$. We suppose that $\mathcal{G}$ contains at least one thermoelastic edge, and we assume that $\mathcal{V}_{ext}$ contains *at least* one element.

© The Author(s), under exclusive license to Springer Nature Switzerland AG 2022
K. Ammari, F. Shel, *Stability of Elastic Multi-Link Structures*, SpringerBriefs in
Mathematics, https://doi.org/10.1007/978-3-030-86351-7_2

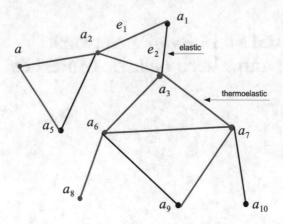

Fig. 2.1 Elastic–thermoelastic network

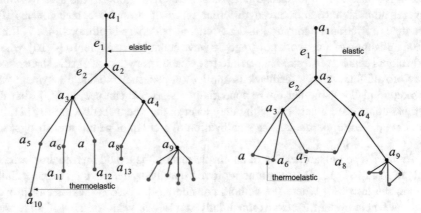

Fig. 2.2 Tree of strings

Let $u_j = u_j(x, t)$ be the function describing the displacement at time t of the body e_j, and in the case where e_j is thermoelastic, we further define $\theta_j = \theta_j(x, t)$ be the temperature difference to a fixed reference temperature of e_j at time t.

Assume the following:

- Every thermoelastic edge e_j satisfies the following equations:

$$u_{j,tt} - u_{j,xx} + \gamma_j \theta_{j,x} = 0 \quad \text{in } (0, \ell_j) \times (0, +\infty), \tag{2.1}$$

$$\theta_{j,t} + \gamma_j u_{j,tx} - \kappa_j \theta_{j,xx} = 0 \quad \text{in } (0, \ell_j) \times (0, +\infty),$$

where γ_j and κ_j are positive constants, with initial conditions

$$u_j(x, 0) = u_j^0(x), \quad u_{j,t}(x, 0) = u_j^1(x), \quad \theta_j(x, 0) = \theta_j^0(x).$$

- Every elastic edge e_j satisfies the following equation:

$$u_{j,tt} - u_{j,xx} = 0 \quad \text{in } (0, \ell_j) \times (0, +\infty),$$

with initial conditions

$$u_j(x, 0) = u_j^0(x), \quad u_{j,t}(x, 0) = u_j^1(x).$$

Denote by J_k^{te} the set of indices of thermoelastic edges incident to a_k and by J_k^e the set of indices of elastic edges incident to a_k. Thus, the boundary conditions on the graph are described as follows:

- The system satisfies the Dirichlet condition for the displacement and temperature at exterior nodes,

$$u_j(a_k, t) = 0, \qquad j \in J_k, \;\; a_k \in \mathcal{V}_{ext},$$
$$\theta_j(a_k, t) = 0, \qquad j \in J_k^{te}, \;\; a_k \in \mathcal{V}_{ext}.$$

- The displacement is continuous at every inner node, that is,

$$u_j(a_k, t) = u_l(a_k, t) \qquad j, l \subset J_k, \;\; a_k \in \mathcal{V}_{int}.$$

- The temperature satisfies the Neumann condition at inner nodes,

$$\theta_{j,x}(a_k, t) = 0, \qquad j \in J_k^{te}, \;\; a_k \in \mathcal{V}_{int}. \tag{2.2}$$

- The system satisfies the balance condition at every inner node,

$$\sum_{j \in J_k^{te}} d_{kj}\left(u_{j,x}(a_k, t) - \gamma_j \theta_j(a_k, t) \right) + \sum_{j \in J_k^e} d_{kj} u_{j,x}(a_k, t) = 0, \;\; a_k \in \mathcal{V}_{int}. \tag{2.3}$$

Note that we can replace the Neumann condition (2.2) for the temperature at all $a_k \in \mathcal{V}_{int}$, with J_k^{te} containing at least two elements, by the continuity condition [48],

$$\theta_j(a_k, t) = \theta_l(a_k, t) = 0,$$

with Kirchhoff's law,

$$\sum_{j \in J_k^{te}} \kappa_j d_{kj} \theta_{j,x}(a_k, t).$$

Then, we get the same results with almost the same proofs.

This chapter is organized as follows: in Sect. 1 we shall formulate our problem in a suitable Hilbert space and deduce the existence and uniqueness of solutions. In Sect. 3 we will prove the exponential stability of the whole system in the case where two elastic edges are not adjacent and in the case of a tree of elastic bodies to which we add in leaves thermoelastic materials.

1 Functional Spaces, Existence, and Uniqueness of Solutions

Denote by J^e the set of indices of the elastic edges and by J^{te} the set of indices of the thermoelastic ones.

For j in J^e, let $\gamma_j = 0$, $\kappa_j = 0$, $V_j = \{0\}$, $V_j^k = \{0\}$, $k = 1, 2$, and for j in J^{te}, let $V_j = L^2(0, \ell_j)$ and $V_j^k = H^k(0, \ell_j)$, $k = 1, 2$.

Thus, the system (2.1)–(2.3) will be equal to the following:

$$u_{j,tt} - u_{j,xx} + \gamma_j \theta_{j,x} = 0, \quad \text{in } (0, \ell_j) \times (0, \infty), \tag{2.4}$$

$$\theta_{j,t} + \gamma_j u_{j,tx} - \kappa_j \theta_{j,xx} = 0, \quad \text{in } (0, \ell_j) \times (0, \infty), \tag{2.5}$$

where $j = 1, \ldots, N$, with conditions

$$u_{j_k}(a_k, t) = 0, \quad \forall a_k \in V_{ext}, \tag{2.6}$$

$$\theta_{j_k}(a_k, t) = 0, \quad \forall a_k \in V_{ext}, \tag{2.7}$$

$$u_j(a_k, t) = u_l(a_k, t), \quad \forall j, l \in J_k, \ \forall a_k \in V_{int}, \tag{2.8}$$

$$\theta_{j,x}(a_k, t) = 0, \quad \forall j \in J_k, \ \forall a_k \in V_{int}, \tag{2.9}$$

$$\sum_{j \in J_k} d_{kj}(u_{j,x}(a_k, t) - \gamma_j \theta_j(a_k, t)) = 0, \quad \forall a_k \in V_{int}, \tag{2.10}$$

and

$$u_j(x, 0) = u_j^0(x), \quad u_{j,t}(x, 0) = u_j^1(x), \quad \theta_j(x, 0) = \theta_j^0(x), \tag{2.11}$$

which is similar to the system (10)–(17) of [1], except that, in the last paper, we consider the continuity condition of the temperature at the interior nodes, instead of the Neumann condition.

Notice that it follows from (2.5) and condition (2.9) that, for all $j \in G_{int}$, we have $\int_0^{\ell_j} (\gamma_j u_{j,xt} + \theta_{j,t}) dx = 0$, that is, $\int_0^{\ell_j} (\gamma_j u_{j,x} + \theta_j) dx$ is conservative all the time. Without loss of generality, we assume that $\int_0^{\ell_j} (\gamma_j u_{j,x} + \theta_j) dx = 0$.

Otherwise, we can make the substitution $\tilde{u}_j = u_j - \frac{\gamma_j c_{0j}^2}{1+\gamma_j^2} x$, $\tilde{v}_j = u_{j,t}$, and $\tilde{\theta}_j = \theta_j - \frac{c_{0j}^2}{1+\gamma_j^2}$, where $c_{0j} = \frac{1}{\ell_j} \int_0^{\ell_j} (\gamma_j (u_j^0)_x + \theta_j^0) dx$.

Denote

$$L^2(\mathcal{G}) = \prod_{j=1}^N L^2(0, \ell_j), \quad H^k(\mathcal{G}) = \prod_{j=1}^N H^k(0, \ell_j), \quad k = 1, 2,$$

$$V(\mathcal{G}) = \prod_{j=1}^N V_j, \quad V^k(\mathcal{G}) = \prod_{j=1}^N V_j^k, \quad k = 1, 2,$$

and set

$$H_0^1(\mathcal{G}) = \left\{ \underline{f} = (f_1, \dots, f_N) \in H^1(\mathcal{G}) \text{ satisfying (2.12) and (2.13)} \right\},$$

$$f_j(a_k) = f_l(a_k), \quad j, l \in J_k, \quad a_k \in V_{int}, \tag{2.12}$$

$$f_{j_k}(a_k) = 0, \quad a_k \in V_{ext}. \tag{2.13}$$

Denote by G_{int} the set of indices of interior edges (i.e., edges with extremities in V_{int}). Then, we define the space

$$\mathcal{H} = \{(\underline{f}, \underline{g}, \underline{h}) \in H_0^1(\mathcal{G}) \times L^2(\mathcal{G}) \times V(\mathcal{G}) \text{ satisfying (2.14)}\},$$

$$\int_0^{\ell_j} (\gamma_j \partial_x f_j + h_j) dx = 0 \text{ for all } j \in G_{int}. \tag{2.14}$$

Equipped with the usual inner product,

$$\langle y, \tilde{y} \rangle_{\mathcal{H}} = \sum_{j=1}^N \left(\left\langle \partial_x f_j, \partial_x \tilde{f}_j \right\rangle + \langle g_j, \tilde{g}_j \rangle + \left\langle h_j, \tilde{h}_j \right\rangle \right)$$

$$= \sum_{j=1}^N \left(\int_0^{\ell_j} \partial_x f_j(x) \partial_x \overline{\tilde{f}}_j(x) dx + \int_0^{\ell_j} g_j(x) \overline{\tilde{g}}_j(x) dx \right.$$

$$\left. + \int_0^{\ell_j} h_j(x) \overline{\tilde{h}}_j(x) dx \right),$$

where $y = (\underline{f}, \underline{g}, \underline{h})$ and $\tilde{y} = (\underline{\tilde{f}}, \underline{\tilde{g}}, \underline{\tilde{h}})$, and the space $\mathcal{H}$ is a Hilbert space.

In this Hilbert space, define the operator $\mathcal{A} : \mathcal{D}(\mathcal{A}) \subseteq \mathcal{H} \to \mathcal{H}$ by

$$
\mathcal{A} = \begin{pmatrix} 0 & A_1 & 0 \\ A_2 & 0 & A_3 \\ 0 & A_3 & A_4 \end{pmatrix}, \tag{2.15}
$$

where $A_1 = diag(I, \dots, I)$, $A_2 = diag(\partial_{xx}, \dots, \partial_{xx})$, $A_3 = diag(-\gamma_1 \partial_x, \dots, -\gamma_N \partial_x)$, $A_4 = diag(\kappa_1 \partial_{xx}, \dots, \kappa_N \partial_{xx})$, and I is the N-unit matrix and with domain

$$
\mathcal{D}(\mathcal{A}) = \left\{ (\underline{u}, \underline{v}, \underline{\theta}) \in \mathcal{H} \cap \left[H^2(\mathcal{G}) \times H_0^1(\mathcal{G}) \times V^2(\mathcal{G}) \right], \text{ satisfying (2.16)} \right\},
$$

where

$$
\begin{cases}
\theta_{j_k}(a_k) = 0, & a_k \in \mathcal{V}_{ext}, \\
\partial_x \theta_j(a_k) = 0, & j \in J_k, \ a_k \in \mathcal{V}_{int}, \\
\sum_{j \in J_k} d_{kj} \left(\partial_x u_j(a_k) - \gamma_j \theta^j(a_k) \right) = 0, & a_k \in \mathcal{V}_{int}.
\end{cases} \tag{2.16}
$$

For $y = (\underline{u}, \underline{v}, \underline{\theta}) \in \mathcal{D}(\mathcal{A})$, the components of $\mathcal{A}y$ are

$$
\begin{cases}
v_j, & j = 1, \dots, N, \\
\partial_{xx} u_j - \gamma_j \partial_x \theta_j, & j = 1, \dots, N, \\
-\gamma_j \partial_x v_j + \kappa_j \partial_{xx} \theta_j, & j = 1, \dots, N.
\end{cases}
$$

Then, the system (2.4)–(2.11) may be rewritten as the first-order evolution equation on $\mathcal{H}$,

$$
\begin{cases}
\dfrac{d}{dt} y = \mathcal{A}y, \\
y(0) = y^0,
\end{cases} \tag{2.17}
$$

where $y = (\underline{u}, \underline{u}_t, \underline{\theta})$ and $y^0 = (\underline{u}^0, \underline{u}^1, \underline{\theta}^0)$.

For $y = (\underline{u}, \underline{v}, \underline{\theta}) \in \mathcal{D}(\mathcal{A})$, a direct calculation gives

$$
Re \langle \mathcal{A}y, y \rangle_{\mathcal{H}} = -\sum_{j=1}^{N} \kappa_j \left\| \partial_x \theta_j \right\|^2 = -\sum_{j \in J^{te}} \kappa_j \left\| \partial_x \theta_j \right\|^2 \leq 0,
$$

which implies that $\mathcal{A}$ is a dissipative operator on $\mathcal{H}$.

Moreover, we have the following result:

Theorem 2.1 *Let $\mathcal{H}$ and $\mathcal{A}$ be defined as before. Then, $1 \in \rho(\mathcal{A})$, the resolvent set of $\mathcal{A}$, and $(\mathcal{I} - \mathcal{A})^{-1}$ is compact.*

Proof Let $z = (\underline{f}, \underline{g}, \underline{h}) \in \mathcal{H}$. We look for an element $y = (\underline{u}, \underline{v}, \underline{\theta}) \in \mathcal{D}(\mathcal{A})$ such that

$$(\mathcal{I} - \mathcal{A})y = z,$$

that is,

$$u_j - v_j = f_j, \quad j = 1, \dots, N, \tag{2.18}$$

$$v_j - \partial_{xx} u_j + \gamma_j \partial_x \theta_j = g_j, \quad j = 1, \dots, N, \tag{2.19}$$

$$\theta_j + \gamma_j \partial_x v_j - \kappa_j \partial_{xx} \theta_j = h_j, \quad j = 1, \dots, N. \tag{2.20}$$

Using (2.18) in (2.19) and (2.20), we have

$$u_j - \partial_{xx} u_j + \gamma_j \partial_x \theta_j = g_j + f_j, \tag{2.21}$$

$$\theta_j + \gamma_j \partial_x u_j - \kappa_j \partial_{xx} \theta_j = h_j + \gamma_j \partial_x f_j, \tag{2.22}$$

for j in $\{1, \dots, N\}$.

We consider the space

$$\mathcal{H}_1 = \{(\underline{w}, \underline{\varphi}) \in H_0^1(\mathcal{G}) \times V^1(\mathcal{G}) \text{ satisfying } (2.23) \text{ and } (2.24)\},$$

$$\varphi_{j_k}(a_k) = 0, \quad a_k \in V_{ext}, \tag{2.23}$$

$$\int_0^{\ell_j} (\gamma_j \partial_x w_j + \varphi_j) dx = 0 \text{ for all } j \in G_{int}, \tag{2.24}$$

equipped with the inner product

$$\langle y, \tilde{y} \rangle_{\mathcal{H}_1} = \sum_{j=1}^N \left(\langle \partial_x w_j, \partial_x \tilde{w}_j \rangle + \langle \varphi_j, \tilde{\varphi}_j \rangle + \kappa_j \langle \partial_x \varphi_j, \partial_x \tilde{\varphi}_j \rangle \right),$$

where $y = (\underline{w}, \underline{\varphi})$ and $\tilde{y} = (\underline{\tilde{w}}, \underline{\tilde{\varphi}})$. Then, $\mathcal{H}_1$ is a Hilbert space.

Let $(\underline{w}, \underline{\varphi})$ in $\mathcal{H}_1$. Multiplying (2.21) by w_j and (2.22) by φ^j for $j = 1, \dots, N$ and summing, we obtain, by taking into account (2.16),

$$a((\underline{u}, \underline{\theta}), (\underline{w}, \underline{\varphi})) = F(\underline{w}, \underline{\varphi}), \tag{2.25}$$

where

$$a((\underline{u}, \underline{\theta}), (\underline{w}, \underline{\varphi})) = \sum_{j=1}^N \left(\int_0^{\ell_j} u_j \overline{w_j} dx + \int_0^{\ell_j} \partial_x u_j \partial_x \overline{w_j} dx - \gamma_j \int_0^{\ell_j} \theta_j \partial_x \overline{w_j} dx \right)$$

$$+ \sum_{j=1}^N \left(\int_0^{\ell_j} \theta_j \overline{\varphi_j} dx + \gamma_j \int_0^{\ell_j} \partial_x u_j \overline{\varphi_j} dx + \kappa_j \int_0^{\ell_j} \partial_x \theta_j \partial_x \overline{\varphi_j} dx \right)$$

$$= \sum_{j=1}^{N} \left(\int_0^{\ell_j} u_j \overline{w_j} dx + \int_0^{\ell_j} \partial_x u_j \partial_x \overline{w_j} dx + \int_0^{\ell_j} \theta_j \overline{\varphi_j} dx \right.$$

$$\left. + \kappa_j \int_0^{\ell_j} \partial_x \theta_j \partial_x \overline{\varphi_j} dx + \gamma_j \left(\int_0^{\ell_j} \partial_x u_j \overline{\varphi_j} dx - \int_0^{\ell_j} \theta_j \partial_x \overline{w_j} dx \right) \right)$$

and

$$F(\underline{w}, \underline{\varphi}) = \sum_{j=1}^{N} \left(\int_0^{\ell_j} (g_j + f_j) \overline{w_j} dx + \int_0^{\ell_j} (h_j + \gamma_j f_j) \overline{w_j} dx \right).$$

It is clear that a is a continuous sesquilinear form on $\mathcal{H}_1 \times \mathcal{H}_1$ and that F is a continuous linear form on $\mathcal{H}_1$. Moreover, for every $(\underline{w}, \underline{\varphi}) \in \mathcal{H}_1$,

$$\left| a(\underline{w}, \underline{\varphi}), (\underline{w}, \underline{\varphi})) \right| \geq \left\| (\underline{w}, \underline{\varphi}) \right\|_{\mathcal{H}_1}^2.$$

Then, by the Lax–Milgram lemma (complex version), (2.25) has a unique solution $(\underline{u}, \underline{\theta}) \in \mathcal{H}_1$.

The classical elliptic theory [51, 57] implies that the solution of (2.21)–(2.22), associated with the conditions

$$u_j(a_k) = u_l(a_k), \quad j, l \in J_k, \quad a_k \in \mathcal{V}_{int},$$

$$\partial_x \theta_j(a_k) = 0, \quad j \in J_k, \quad a_k \in \mathcal{V}_{int},$$

$$u_{j_k}(a_k) = 0, \quad \theta_{j_k}(a_k) = 0, \quad a_k \in \mathcal{V}_{ext},$$

$$\int_0^{\ell_j} (\gamma_j \partial_x f_j + h_j) dx = 0 \quad \text{for all } j \in G_{int},$$

$$\sum_{j \in J_k} d_{kj} \left(\partial_x u_j(a_k) - \gamma_j \theta_j(a_k) \right) = 0, \quad a_k \in \mathcal{V}_{int},$$

belongs to the space $H^2(\mathcal{G}) \times V^2(\mathcal{G})$, and

$$\|y\|_{\mathcal{H}}^2 \leq c \|z\|_{\mathcal{H}}^2,$$

where c is a positive constant independent of y, which proves that $(\underline{u}, \underline{v}, \underline{\theta}) \in \mathcal{D}(\mathcal{A})$ and $(\mathcal{I} - \mathcal{A})^{-1} \in \mathcal{L}(\mathcal{H})$, that is, $1 \in \rho(\mathcal{A})$.

The Sobolev embedding theorem asserts that $(\mathcal{I} - \mathcal{A})^{-1}$ is a compact operator. This finishes the proof of the theorem.   □

Since $1 \in \rho(\mathcal{A})$ and $\mathcal{A}$ is dissipative, we have the following result due to Theorem 1.15:

Corollary 2.2 *The operator $\mathcal{A}$ generates a C_0-semigroup of contraction $(S(t))_{t \geq 0}$ on the Hilbert space $\mathcal{H}$.*

Therefore, for an initial datum $y^0 \in \mathcal{H}$, there exists a unique solution

$$y \in C([0, +\infty), \mathcal{H})$$

of the Cauchy problem (2.17).

Moreover, if $y^0 \in \mathcal{D}(\mathcal{A})$, then

$$y \in C([0, +\infty), \mathcal{D}(\mathcal{A})) \cap C^1([0, +\infty), \mathcal{H}).$$

2 Exponential Stability

In this section, we study the asymptotic behavior of the system (2.4)–(2.11). Precisely, we will prove that the C_0-semigroup $(S(t))_{t \geq 0}$ is exponentially stable.

Recall that a C_0-semigroup $(T(t))_{t \geq 0}$ is exponentially stable if and only if there exist constants $C \geq 1$ and $a > 0$ such that

$$\|T(t)\| \leq Ce^{-at}, \quad \forall t \geq 0.$$

We will use the frequency domain condition due to Gearhard (Theorem 1.25).

2.1 First Case

In this section, we consider the case where two elastic edges are not adjacent (Fig. 2.1).

Lemma 2.3 *Let $\mathcal{A}$ be the operator given by (2.15), then condition (1.9) holds for $(S(t))_{t \geq 0}$, that is, $i\mathbb{R} = \{i\beta \mid \beta \in \mathbb{R}\} \subseteq \rho(\mathcal{A})$. In particular, by Theorem 1.22, the semigroup $(S(t))_{t \geq 0}$ is strongly stable.*

Proof Suppose that (1.9) is not true. Then, there is a real number $\beta \in \mathbb{R}$ such that $\lambda := i\beta \in \sigma(\mathcal{A})$, the spectrum of $\mathcal{A}$. By the fact that $(\mathcal{I} - \mathcal{A})^{-1}$ is compact, the spectrum of $\mathcal{A}$ consists of all isolated eigenvalues, that is, $\sigma(\mathcal{A}) = \sigma_p(\mathcal{A})$, and then there exists $y = (\underline{u}, \underline{v}, \theta) \in \mathcal{D}(\mathcal{A})$, $y \neq 0$, such that $\mathcal{A}y = i\beta y$.

Then,

$$Re(\langle \mathcal{A}y, y \rangle_{\mathcal{H}}) = Re(\langle \lambda y, y \rangle_{\mathcal{H}}) = 0.$$

This leads to

$$-\sum_{j=1}^{N} \kappa_j \left\| \partial_x \theta_j \right\|^2 = 0,$$

which implies that $\partial_x \theta_j = 0$ for $j = 1, \ldots, N$.

Then, θ_j is constant for $j = 1, \ldots, N$, and $(\underline{u}, \underline{v})$ satisfies

$$\begin{cases} v_j = \lambda u_j, & j = 1, \ldots, N, \\ \partial_{xx} u_j - \gamma_j \partial_x \theta_j = \lambda^2 u_j, & j = 1, \ldots, N, \\ -\lambda \gamma_j \partial_x u_j = \lambda \theta_j, & j = 1, \ldots, N. \end{cases} \tag{2.26}$$

If $\lambda = 0$, then $v_j = 0$ and $\partial_x u_j$ is constant for $j = 1, \ldots, N$. Multiplying the second equation in (2.26) by u_j and summing, we obtain

$$\sum_{j=1}^{N} \left(-\int_0^{\ell_j} |\partial_x u_j|^2 \, dx + \gamma_j \int_0^{\ell_j} \theta_j \partial_x \overline{u_j} dx \right) = 0.$$

Then, using (2.14) and the fact that θ_j and $\partial_x u_j$ are constant for $j = 1, \ldots, N$, we deduce that $\partial_x \underline{u} = 0$ and $\underline{\theta} = 0$.

Now, suppose that $\lambda \neq 0$. For j in J^{te}, the third and second equations of (2.26) imply $u_j = 0$, and then $\partial_x u_j = 0$, $\theta_j = 0$, and $v_j = 0$ in $L^2(0, \ell_j)$. Moreover, u_j, $\partial_x u_j$, and θ_j vanish on both ends of e_j. For j in J^e, by taking into account that $u \in H_0^1(\mathcal{G})$, condition (2.16), and the fact that two elastic edges are not adjacent, u_j satisfies the Cauchy problem $\partial_{xx} u_j = \lambda^2 u_j$, $u_j(a_k) = 0$, and $\partial_x u_j(a_k) = 0$, where a_k is an end of e_j, then $u_j = 0$; hence, by the first equation of (2.26), $v_j = 0$ in $L^2(0, \ell_j)$, which contradicts the fact that $y = (\underline{u}, \underline{v}, \underline{\theta})$ is an eigenvector of $\mathcal{A}$. We conclude that $i\mathbb{R} \subset \rho(\mathcal{A})$. $\qquad\square$

Lemma 2.4 *Let $\mathcal{A}$ be the operator given by (2.15), then condition (1.10) holds.*

Proof Suppose that (1.10) is not true, then there exists a sequence (β_n) of real numbers, with $\beta_n \longrightarrow \infty$ (without loss of generality, we suppose that $\beta_n > 0$), and a sequence of vectors $(y_n) = (\underline{u}_n, \underline{v}_n, \underline{\theta}_n)$ in $\mathcal{D}(\mathcal{A})$ with $\|y_n\|_{\mathcal{H}} = 1$, such that

$$\|(i\beta_n - \mathcal{A})y_n\| \longrightarrow 0. \tag{2.27}$$

Our goal is to prove that this condition yields the contradiction $\|y_n\|_{\mathcal{H}} \longrightarrow 0$ as $n \longrightarrow 0$. The proof is divided into two steps. In the first, we prove, as in the proof of Theorem 2 in [1], that $\partial_x u_{j,n}$, $\theta_{j,n}$, and $v_{j,n}$ converge to 0 in $L^2(0, \ell_j)$ for j in J^{te}. In the second, we prove the same properties for j in J^e.

First Step Because $Re(\langle(i\beta_n - \mathcal{A})y_n, y_n\rangle_{\mathcal{H}}) = \sum_{j\in J^{te}} \kappa_j \left\|\partial_x\theta_{j,n}\right\|^2$, we obtain

$$\left\|\partial_x\theta_{j,n}\right\| \longrightarrow 0. \tag{2.28}$$

Now, writing condition (2.27) term by term, we obtain, for all j in $\{1, \ldots, N\}$,

$$i\beta_n u_{j,n} - v_{j,n} \longrightarrow 0, \quad \text{in } H^1(0, \ell_j), \tag{2.29}$$

$$i\beta_n v_{j,n} - \partial_{xx}u_{j,n} + \gamma_j\partial_x\theta_{j,n} \longrightarrow 0, \quad \text{in } L^2(0, \ell_j), \tag{2.30}$$

$$i\beta_n\theta_{j,n} + \gamma_j\partial_x v_{j,n} - \kappa_j\partial_{xx}\theta_{j,n} \longrightarrow 0, \quad \text{in } L^2(0, \ell_j). \tag{2.31}$$

Let $j \in J^{te}$.
Multiplying (2.29) by γ_j and adding the result up to (2.31), we obtain

$$i\beta_n(\gamma_j\partial_x u_{j,n} + \theta_{j,n}) - \kappa_j\partial_{xx}\theta_{j,n} \longrightarrow 0 \text{ in } L^2(0, \ell_j). \tag{2.32}$$

Taking L^2-inner product of (2.32) with $\dfrac{\gamma_j\partial_x u_{j,n} + \theta_{j,n}}{\beta_n}$, which is bounded in $L^2(0, \ell_j)$, and integrating by parts, we obtain

$$i\left\|\gamma_j\partial_x u_{j,n} + \theta_{j,n}\right\|^2 + \left\langle\frac{\kappa_j}{\beta_n}\partial_x\theta_{j,n}, \gamma_j\partial_{xx}u_{j,n} + \partial_x\theta_{j,n}\right\rangle$$

$$- \frac{\kappa_j\partial_x\theta_{j,n}(\gamma_j\partial_x\overline{u}_{j,n} + \overline{\theta}_{j,n})}{\beta_n}\Bigg|_{x=0}^{x=\ell_j} \longrightarrow 0. \tag{2.33}$$

By (2.28) and the fact that $\left\|\dfrac{\partial_{xx}u_{j,n}}{\beta_n}\right\|$ is bounded, we can conclude that the product in the aforementioned expression tends to 0.

Recall the Gagliardo–Nirenberg inequality (1.13) of Theorem 1.33: there are two positive constants C_1 and C_2 such that, for any w in $H^1(0, \ell_j)$,

$$\|w\|_{L^\infty} \leq C_1\left\|\partial_x w\right\|^{1/2}\|w\|^{1/2} + C_2\|w\|.$$

With this inequality applied to $w = \dfrac{\partial_x\theta_{j,n}}{\sqrt{\beta_n}}$, $w = \dfrac{\partial_x u_{j,n}}{\sqrt{\beta_n}}$ and $\dfrac{\theta_{j,n}}{\sqrt{\beta_n}}$ yield that the boundary terms in (2.33) converge to 0.
Then,

$$\gamma_j\partial_x u_{j,n} + \theta_{j,n} \longrightarrow 0 \text{ in } L^2(0, \ell_j). \tag{2.34}$$

It follows from (2.34) and (2.28) that $\theta_{j,n}$ is uniformly bounded in $H^1(0, \ell_j)$. Then, by the compactness of the embedding of $H^1(0, \ell_j)$ into $L^2(0, \ell_j)$, there is a subsequence of $\theta_{j,n}$, still denoted $\theta_{j,n}$, which is a Cauchy sequence in $L^2(0, \ell_j)$. Then, with (2.28) used again, $\theta_{j,n}$ is a Cauchy sequence in $H^1(0, \ell_j)$. Let θ^j be its limit. Then, from (2.29) and (2.34), it follows that $u_{j,n}$ converges to 0 in $H^1(0, \ell_j)$ and $\theta_j = 0$.

Dividing (2.30) by β_n, simplifying by taking into account that $\partial_x \theta_{j,n} \longrightarrow 0$ in $L^2(0, \ell_j)$, multiplying the result by $v_{j,n}$, and integrating by parts, we obtain

$$\mathbf{i} \left\| v_{j,n} \right\|^2 + \frac{1}{\beta_n} \left\langle \partial_x v_{j,n}, \partial_x u_{j,n} \right\rangle - \frac{v_{j,n} \partial_x \overline{u}_{j,n}}{\beta_n} \bigg|_{x=0}^{x=\ell_j} \longrightarrow 0. \tag{2.35}$$

As for (2.33), we prove that the second and third terms in (2.35) converge to 0, so $v_{j,n} \longrightarrow 0$ in $L^2(0, \ell_j)$.

Second Step We will prove that $\partial_x u_{j,n}$ and $v_{j,n}$ converge to 0 in $L^2(0, \ell_j)$ for j in J^e.

Let j in $\{1, \ldots, N\}$.

Equations (2.29) and (2.30) can be rewritten, respectively, as

$$\mathbf{i} \beta_n u_{j,n} - v_{j,n} = f_{j,n} \longrightarrow 0 \quad \text{in } H^1(0, \ell_j), \tag{2.36}$$

$$\mathbf{i} \beta_n v_{j,n} - \partial_{xx} u_{j,n} = g_{j,n} \longrightarrow 0 \quad \text{in } L^2(0, \ell_j). \tag{2.37}$$

Substituting (2.36) into (2.37), we obtain

$$\partial_{xx} u_{j,n} + \beta_n^2 u_{j,n} = -g_n - \mathbf{i} \beta_n f_{j,n}. \tag{2.38}$$

From (2.36), we deduce immediately that

$$\beta_n^2 \left\| u_{j,n} \right\|^2 - \left\| v_{j,n} \right\|^2 \longrightarrow 0. \tag{2.39}$$

Next, let h_j be a function in $C^1([0, \ell_j], \mathbb{C})$. We want to compute the real part of the inner product of (2.38) with $h_j \partial_x u_{j,n}$ in $L^2(0, \ell_j)$.

It follows from integration by parts that

$$Re\left(\left\langle \partial_{xx} u_{j,n}, h_j \partial_x u_{j,n} \right\rangle \right) = \frac{1}{2} \left(\left| \partial_x u_{j,n}(x) \right|^2 h_j(x) \bigg|_{x=0}^{x=\ell_j} \right.$$

$$\left. - \int_0^{\ell_j} \left(\left| \partial_x u_{j,n}(x) \right|^2 \right) \partial_x h_j(x) dx \right),$$

$$Re\left(\left\langle \beta_n^2 u_{j,n}, h_j \partial_x u_{j,n} \right\rangle \right) = \frac{1}{2} \left(\beta_n^2 \left| u_{j,n}(x) \right|^2 h_j(x) \bigg|_{x=0}^{x=\ell_j} \right.$$

$$-\int_0^{\ell_j} \beta_n^2 \left|u_{j,n}(x)\right|^2 \partial_x h_j(x)dx\bigg),$$

and

$$\langle i\beta_n f_{j,n} + g_{j,n}, h_j \partial_x u_{j,n}\rangle = i\beta_n f_{j,n}(x)h_j(x)\overline{u_{j,n}}(x)\big|_{x=0}^{x=\ell_j} + \langle \partial_x f_{j,n}, i\beta_n u_{j,n} h_j\rangle$$
$$+ \langle f_{j,n}, i\beta_n u_{j,n} \partial_x h_j\rangle + \langle g_{j,n}, h_j \partial_x u_{j,n}\rangle. \qquad (2.40)$$

By taking into account that $\|g_{j,n}\|$, $\|\partial_x f_{j,n}\|$, and $\|f_{j,n}\|$ converge to 0 and that $\|\partial_x u_{j,n}\|$ and $\|i\beta_n u_{j,n}\|$ are bounded, we deduce that the products in the second member of (2.40) converge to zero.

Hence, the real part of the inner product of (2.38) by $h_j \partial_x u_{j,n}$ gives

$$\frac{1}{2}\beta_n^2 \left|u_{j,n}(x)\right|^2 h_j(x)\Big|_{x=0}^{x=\ell_j} + \frac{1}{2}\left|\partial_x u_{j,n}(x)\right|^2 h_j(x)\Big|_{x=0}^{x=\ell_j}$$

$$-\frac{1}{2}\int_0^{\ell_j}\left(\left|\partial_x u_{j,n}(x)\right|^2 + \beta_n^2 \left|u_{j,n}(x)\right|^2\right)\partial_x h_j(x)dx$$

$$+ Re\left(i\beta_n f_{j,n}(x)h_j(x)\overline{u_{j,n}}(x)\right)\Big|_{x=0}^{x=\ell_j} \longrightarrow 0. \qquad (2.41)$$

For j in J^{te}, with $h_j(x) = \ell_j - x$, and by taking into account that $u_{j,n} \longrightarrow 0$ in $H^1(0, \ell_j)$ and $\beta_n u_{j,n} = -i(f_{j,n} + v_{j,n}) \longrightarrow 0$ in $L^2(0, \ell_j)$, we obtain

$$-\frac{1}{2}\beta_n^2 \left|u_{j,n}(0)\right|^2 - \frac{1}{2}\left|\partial_x u_{j,n}(0)\right|^2 - Re\left(i\beta_n f_{j,n}(0)\overline{u_{j,n}}(0)\right) \longrightarrow 0. \qquad (2.42)$$

Now, because

$$-Re\left(i\beta_n f_{j,n}(0)\overline{u_{j,n}}(0)\right) \leq \frac{1}{4}\beta_n^2 \left|u_{j,n}(0)\right|^2 + \left|f_{j,n}(0)\right|^2,$$

we can deduce

$$-\frac{1}{2}\beta_n^2 \left|u_{j,n}(0)\right|^2 - \frac{1}{2}\left|\partial_x u_{j,n}(0)\right|^2 - Re\left(i\beta_n f_{j,n}(0)\overline{u_{j,n}}(0)\right) - \left|f_{j,n}(0)\right|^2$$

$$\leq -\frac{1}{4}\beta_n^2 \left|u_{j,n}(0)\right|^2 - \frac{1}{2}\left|\partial_x u_{j,n}(0)\right|^2 \leq 0. \qquad (2.43)$$

The continuity condition (2.12) and the Dirichlet condition (2.13) of $\underline{f}_n$ imply that $f_{j,n}(0) \longrightarrow 0$.

Then, by (2.42), the first member of the inequalities in (2.43) converges to 0; hence, the second member converges to 0 and then

$$\beta_n^2 \left| u_{j,n}(0) \right|^2 \to 0 \quad \text{and} \quad \left| \partial_x u_{j,n}(0) \right|^2 \to 0, \tag{2.44}$$

which implies, with the use of (2.42),

$$Re\left(i\beta_n f_{j,n}(0)\overline{u_{j,n}}(0) \right) \to 0. \tag{2.45}$$

By the same manner, if we choose $h_j(x) = x$, we deduce that

$$\beta_n^2 \left| u_{j,n}(\ell_j) \right|^2 \to 0, \quad \left| \partial_x u_{j,n}(\ell_j) \right|^2 \to 0, \tag{2.46}$$

and

$$Re\left(i\beta_n f_{j,n}(\ell_j)\overline{u_{j,n}}(\ell_j) \right) \to 0. \tag{2.47}$$

Now, let j in J^e, and let a_k be an end of e_j in $\mathcal{V}_{int}$.

By the fact that $\underline{u}_n$ is continuous at the inner nodes and using (2.44) and (2.45) or (2.46) and (2.47), we deduce that

$$\beta_n^2 \left| u_{j,n}(a_k) \right|^2 \to 0 \quad \text{and} \quad Re\left(i\beta_n f_{j,n}(a_k)\overline{u_{j,n}}(a_k) \right) \to 0.$$

On the other hand, because $\underline{u}_n$ and $\underline{\theta}_n$ satisfy the balance condition in (2.16) and $\underline{\theta}_n$ converges to 0 at every inner node, then (2.44) or (2.46) implies

$$\left| \partial_x u_{j,n}(a_k) \right|^2 \to 0.$$

Hence, from (2.41) with $h_j(x) = \ell_j - x$ or $h_j(x) = x$, we obtain

$$\int_0^{\ell_j} \left(\left| \partial_x u_{j,n}(x) \right|^2 + \beta_n^2 \left| u_{j,n}(x) \right|^2 \right) dx \longrightarrow 0,$$

which concludes that $\| y_n \|_{\mathcal{H}} \longrightarrow 0$. This behavior contradicts the hypothesis that y_n has the unit norm. □

We can now state the main result of this chapter.

Theorem 2.5 *The C_0-semigroup $S(t)$, generated by the operator $\mathcal{A}$, is exponentially stable.*

Proof The proof is a direct consequence of Lemmas 2.3 and 2.4. □

Remark 2.6 Consider the case where the thermoelastic edges are governed by Cattaneo's law.

The linear system on the thermoelastic edge e_j will be

$$u_{j,tt} - \alpha_j u_{j,xx} + \beta_j \theta_{j,x} = 0, \quad \text{in } (0, \ell_j) \times (0, \infty),$$

$$\theta_{j,t} + \gamma_j q_{j,x} + \delta_j u_{j,tx} = 0, \quad \text{in } (0, \ell_j) \times (0, \infty),$$

$$\tau_j q_{j,t} + q_j + \kappa_j \theta_{j,x} = 0, \quad \text{in } (0, \ell_j) \times (0, \infty),$$

where $q_j = q_j(x, t)$ is the heat flux at time t of the edge e_j and $\alpha_j, \beta_j, \gamma_j, \delta_j, \tau_j$, and κ_j are positive constants, with initial conditions

$$u_j(x, 0) = u_j^0(x), \quad u_{j,t}(x, 0) = u_j^1(x), \quad \theta_j(x, 0) = \theta_j^0(x), \quad q_j(x, 0) = q_j^0(x).$$

Every elastic edge satisfies the following equation:

$$u_{j,tt} - \alpha_j u_{j,xx} = 0 \quad \text{in } (0, \ell_j) \times (0, +\infty),$$

where α_j is a positive constant, with initial conditions

$$u_j(x, 0) = u_j^0(x), \quad u_{j,t}(x, 0) = u_j^1(x).$$

In addition, suppose that the whole system will satisfy the following boundary conditions:

$$u_j(a_k, t) = 0, \qquad j \in J_k, \quad a_k \in \mathcal{V}_{ext},$$

$$\theta_j(a_k, t) = 0, \qquad j \in J_k^{te}, \quad a_k \in \mathcal{V}_{ext},$$

$$u_j(a_k, t) = u_l(a_k, t) \qquad j, l \in J_k, \quad a_k \in \mathcal{V}_{int},$$

$$q_j(a_k, t) = 0 \qquad j \in J_k^{te}, \quad a_k \in \mathcal{V}_{int},$$

$$\sum_{j \in J_k^{te}} d_{kj} \frac{\delta_j}{\beta_j} \left(\alpha_j \partial_x u_j(a_k, t) - \beta_j \theta_j(a_k, t) \right) + \sum_{j \in J_k^e} d_{kj} \alpha_j \partial_x u_j(a_k, t) = 0, \quad a_k \in \mathcal{V}_{int}.$$

Then, we can also prove that solutions of the system decay exponentially.

2.2 Second Case

Now, we consider the second case where the graph is a tree of elastic edges that ends by thermoelastic edges at leaves (Fig. 2.2). We can prove that the system is exponentially stable.

Theorem 2.7 *The C_0-semigroup, generated by the operator A, is exponentially stable.*

Proof With the frequency domain condition, we only need to justify (1.9) and (1.10). The proofs are similar to those of Lemmas 2.3 and 2.4, respectively, with some modifications.

Suppose that (1.9) is not true; as in the proof of Lemma 2.3, there are a real number β, $\beta \neq 0$, and a vector $y = (u, v, \theta) \in \mathcal{D}(\mathcal{A})$, $y \neq 0$, such that $\mathcal{A}y = i\beta y$. If e_j is thermoelastic, then the functions u_j, $\partial_x u_j$, and θ_j are zero in $L^2(0, \ell_j)$ and at the inner nodes; thus, (u, v) satisfies

$$\begin{cases} v_j = \lambda u_j, & j = 1, \dots, N, \\ \partial_{xx} u_j = \lambda^2 u_j, & j = 1, \dots, N. \end{cases} \tag{2.48}$$

Now, if e_j is an elastic edge attached to a thermoelastic one, then it satisfies the Cauchy problem $\partial_{xx} u_j = \lambda^2 u_j$, $u_j(a_k) = 0$, and $\partial_x u_j(a_k) = 0$, where a_k is an end of e_j, and then $u_j = 0$ in $H^1(0, \ell_j)$ and $v_j = 0$ in $L^2(0, \ell_j)$; moreover, u_j and $\partial_x u_j$ vanish at the ends of e_j. We iterate this procedure from the leaves to the root, so that we obtain $\underline{u} = 0$ in $H_0^1(\mathcal{G})$, $\underline{v} = 0$ in $L^2(\mathcal{G})$, and $\underline{\theta} = 0$ in $V(\mathcal{G})$, which contradict the fact that $y \neq 0$.

For property (1.10), the difference from the proof of Lemma 2.4 is that, in the second step of such proof, we prove first that $\int_0^{\ell_j} \left(\left| \partial_x u_{j,n}(x) \right|^2 + \left| v_{j,n}(x) \right|^2 \right) dx$ converges to zero when e_j is an elastic edge attached to a thermoelastic one, and then, by iteration using the same arguments, we prove that $\int_0^{\ell_j} \left(\left| \partial_x u_{j,n}(x) \right|^2 + \left| v_{j,n}(x) \right|^2 \right) dx$ converges to zero for every elastic edge, which contradicts the fact that $\| y_n \| = 1$.   □

3 Comments

3.1 Comment 1

We can consider a more general case (as in the next chapter): we assume that $\mathcal{G}$ contains at least one thermoelastic string, that $\mathcal{V}_{ext} \neq \emptyset$, that every maximal subgraph of elastic edges is a tree, the leaves of which thermoelastic edges are attached, and that every maximal subgraph of thermoelastic edges is not a circuit. Then, we can prove that the associated semigroup is exponentially stable.

3.2 Comment 2

In [37], $\mathcal{G}$ is a star-shaped network of interconnected elastic and thermoelastic rods (Fig. 2.3). The edges (rods) e_j, $j = 1, \dots, N$ occupy the intervals $(0, \ell_j)$, $\ell_j > 0$,

Fig. 2.3 Star-shaped
thermoelastic network

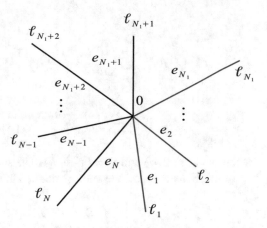

respectively. The common node is identified to $x = 0$. The edges $e_j, \; j = 1, \ldots, N_1$
$(0 < N_1 < N)$ are thermoelastic, and the other edges are all purely elastic.

The thermoelastic–elastic network system under consideration is

$$
\begin{cases}
u_{j,tt}(x,t) - u_{j,xx}(x,t) + \alpha_j \theta_{j,x}(x,t) = 0, \; x \in (0, \ell_j), \; j = 1, 2, \ldots, N_1, \; t > 0, \\
\theta_{j,t}(x,t) - \theta_{j,xx}(x,t) + \beta_j u_{j,tx}(x,t) = 0, \; x \in (0, \ell_j), \; j = 1, 2, \ldots, N_1, \; t > 0, \\
u_{j,tt}(x,t) - u_{j,xx}(x,t) = 0, \; x \in (0, \ell_j), \; j = N_1 + 1, N_1 + 2, \ldots, N, \; t > 0, \\
u_j(\ell_j, t) = 0, \; j = 1, 2, \ldots, N, \; t > 0, \\
u_j(0, t) = u_k(0, t), \; \forall\, j, k = 1, 2, \ldots, N, \; t > 0, \\
\theta_k(\ell_k, t) = 0, \; k = 1, 2, \ldots, N_1, \; t > 0, \\
\theta_j(0, t) = \theta_k(0, t), \; \forall\, j, k = 1, 2, \ldots, N_1, \; t > 0, \\
\sum_{j=1}^{N} u_{j,x}(0, t) = \sum_{j=1}^{N_1} \alpha_j \theta_j(0, t), \; \sum_{j=1}^{N_1} \frac{\alpha_j}{\beta_j} \theta_{j,x}(0, t) = 0, \; t > 0, \\
\theta^k(t = 0) = \theta_k^0, \; k = 1, 2, \ldots, N_1, \\
u_j(t = 0) = u_j^0, \; u_{j,t}(t = 0) = u_j^1, \; j = N_1 + 1, N_1 + 2, \ldots, N,
\end{cases}
$$

(2.49)

where $\left(\left(u_j^0 \right)_{j=N_1+1}^{N}, \left(u_j^1 \right)_{j=N_1+1}^{N}, \left(\theta_k^0 \right)_{k=1}^{N_1} \right)$ is the given initial state.

The natural energy of this system is as follows:

$$
E(t) = \frac{1}{2} \sum_{j=1}^{N} \int_0^{\ell_j} \left[u_{j,t}^2 + u_{j,x}^2 \right] dx + \frac{1}{2} \sum_{j=1}^{N_1} \frac{\alpha_j}{\beta_j} \int_0^{\ell_j} \theta_j^2 dx.
$$

The space state is

$$
\mathcal{H} = H_0^1(\mathcal{G}) \times L^2(\mathcal{G}) \times \left(\prod_{k=1}^{N_1} L^2(0, \ell_k) \right),
$$

equipped with the inner product

$$\langle y, \tilde{y}\rangle_{\mathcal{H}} = \sum_{j=1}^{N} \left(\int_0^{\ell_j} \partial_x f_j(x) \partial_x \overline{\tilde{f}_j}(x) dx + \int_0^{\ell_j} g_j(x) \overline{\tilde{g}_j}(x) dx \right.$$

$$\left. + \sum_{k=1}^{N_1} \int_0^{\ell_k} \frac{\alpha_k}{\beta_k} h_k(x) \overline{\tilde{h}_k}(x) dx \right),$$

where $y = (\underline{f}, \underline{g}, \underline{h})$ and $\tilde{y} = (\underline{\tilde{f}}, \underline{\tilde{g}}, \underline{\tilde{h}})$. The space $\mathcal{H}$ is a Hilbert space.

In this Hilbert space, the authors defined the operator $\mathcal{A} : \mathcal{D}(\mathcal{A}) \subseteq \mathcal{H} \to \mathcal{H}$, by

$$\mathcal{A} = \begin{pmatrix} 0 & I & 0 \\ \partial_{xx} & 0 & -\alpha I_{N \times N_1} \partial_x \\ 0 & -\beta I_{N \times N_1}^T \partial_x & \partial_{xx} \end{pmatrix}, \tag{2.50}$$

where $\alpha = diag(\alpha_1, \alpha_2, \ldots, \alpha_{N_1})$, $\beta = diag(\beta_1, \beta_2, \ldots, \beta_{N_1})$, $I_{N \times N_1} = [I_{N_1,0}]^T$, and I_{N_1} is the N_1-unit matrix, with domain

$$\mathcal{D}(\mathcal{A}) = \left\{ (\underline{u}, \underline{v}, \underline{\theta}) \in \left[H_0^1 \cap \prod_{j=1}^{N} H^2(0, \ell_j) \right] \times H_0^1(\mathcal{G}) \right.$$

$$\left. \times \prod_{j=1}^{N_1} H^2(0, \ell_j), \text{ satisfying } (2.51) \right\},$$

where

$$\begin{cases} \sum_{j=1}^{N} \partial_x u_j(0) = \sum_{j=1}^{N_1} \alpha_j \partial_x \theta_j(0), \\ \theta_j(\ell_j) = 0, \quad j = 1, 2 \ldots, N_1, \\ \theta_j(0) = \theta_k(0), \quad j, k = 1, 2, \ldots, N_1, \\ \sum_{j=1}^{N_1} \frac{\alpha_j}{\beta_j} \partial_x \theta_j(0) = 0. \end{cases} \tag{2.51}$$

Then, the system (2.49) may be rewritten as the first-order evolution equation on $\mathcal{H}$,

$$\begin{cases} \dfrac{d}{dt} y = \mathcal{A} y, \\ y(0) = y^0, \end{cases}$$

where $y = (\underline{u}, \underline{u}_t, \underline{\theta})$ and $y^0 = (\underline{u}^0, \underline{u}^1, \underline{\theta}^0)$.

The authors proved the following results:

Theorem 2.8 *Operator $\mathcal{A}$ generates a C_0-semigroup of contractions on $\mathcal{H}$. Moreover, the energy of the system (2.49) decays to zero as $t \to \infty$ if and only if one of the following two conditions is fulfilled:*

(1) $N - N_1 = 1$ *and*
(2) $N - N_1 \geq 2$ *and $\ell_i / \ell_j \notin \mathbb{Q}$, $i, j = N_1 + 1, N_1 + 2, \ldots, N$, $i \neq j$.*

Theorem 2.9 *The energy of system (2.49) decays to zero exponentially if and only if $N - N_1 \leq 1$, that is, if no more than one purely elastic undamped rod is involved in the network.*

To examine more closely the case $N - N_1 > 1$, we need the following definition:

Definition 2.10 ([18, 85]) Real numbers $\ell_1, \ell_2, \ldots, \ell_m$ are said to verify the conditions (S), if $\ell_1, \ell_2, \ldots, \ell_m$ are linearly independent over the field $\mathbb{Q}$ of rational numbers, and the ratios ℓ_i / ℓ_j are algebraic numbers for $i, j = 1, 2, \ldots, m$.

Theorem 2.11 *When $N - N_1 > 1$, one has*

$$\liminf_{t \to \infty} t E(t) > 0. \tag{2.52}$$

Thus, we cannot expect a decay rate which is beyond first-order polynomial. Furthermore, if $\ell_{N_1+1}, \ell_{N_1+2}, \ldots, \ell_N$ $(N - N_1 > 1)$ satisfy the condition (S), then for any $\varepsilon > 0$, there always exists a constant $C_\varepsilon > 0$ such that the energy of network (2.49) satisfies

$$E(t) \leq C_\varepsilon t^{-\frac{1}{1+\varepsilon}} \| (\underline{u}^0, \underline{u}^1, \underline{\theta}^0) \|^2_{\mathcal{D}(\mathcal{A})}, \ \forall t \geq 0$$

for all $(\underline{u}^0, \underline{u}^1, \underline{\theta}^0) \in \mathcal{D}(\mathcal{A})$. Thus, this decay rate is nearly sharp in the sense of (2.52).

Chapter 3
Exponential Stability of a Network of Beams

In this chapter, we consider the asymptotic behavior of a network of Euler–Bernoulli beams, some of them are thermoelastic (sensible to a thermal effect), while the others are elastic (without temperature change). Such study is motivated by the need for engineers to eliminate vibrations in some dynamical structures consisting of elastic beams, coupled in the form of chain or graph such as pipelines, bridges, and some cable networks. There are other complicated examples in the automotive industry.

Our main question is whether dissipation over each thermoelastic edge is enough to produce an exponential decay rate of the whole system. As in [1, 88], where we considered a network of strings, we prove by using the frequency domain method that the answer is positive. The main difficulty in the present case is to "estimate" some boundary terms of higher order. To overcome it, we improve the method used in [88] by special multiplier techniques as in [5].

Many authors have studied similar networks, for example, Mercier and Regnier in [68, 69] studied the controllability and the spectrum of a network of Euler–Bernoulli beams. Dekoninck and Nicaise studied in [21] the exact controllability of a network of beams with boundary dampings. Ammari in [3] has proved the polynomial stability of a star-shaped network of elastic Euler–Bernoulli beams. In [100], the authors proved the exponential stability of a star-shaped network of beams with controls applied at external ends. In [35] the authors have proved the asymptotic stability of a star-shaped network of Timoshenko Beams. Ammari et al. considered in [12] a chain of Euler–Bernoulli beams and strings; they established some results of polynomial stability of such system. For the transmission problem between elastic and thermoelastic materials see, for example, [62, 81] for strings, and [84] for beams.

Suppose that the equilibrium position of our network of elastic and thermoelastic beams coincides with the graph $\mathcal{G}$ of N edges, $e_1, \ldots, e_N$, and p vertices, $a_1, \ldots, a_p$. We assume that $\mathcal{G}$ contains at least one thermoelastic beam, that $\mathcal{V}_{ext} \neq \emptyset$, that every maximal subgraph of elastic edges is a tree, the leaves of which

© The Author(s), under exclusive license to Springer Nature Switzerland AG 2022
K. Ammari, F. Shel, *Stability of Elastic Multi-Link Structures*, SpringerBriefs in
Mathematics, https://doi.org/10.1007/978-3-030-86351-7_3

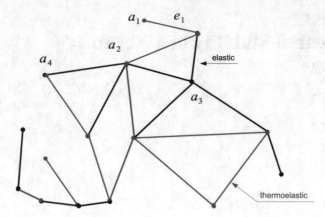

Fig. 3.1 Elastic–thermoelastic graph

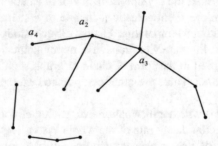

Fig. 3.2 In Fig. 3.1, there are three maximal subgraphs of elastic edges, each of them is a tree

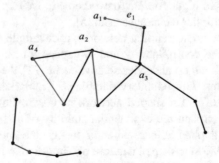

Fig. 3.3 Some thermoelastic edges attached to every leaf of the trees of Fig. 3.2

thermoelastic edges are attached, and that every maximal subgraph of thermoelastic edges is not a circuit (Figs. 3.1, 3.2, and 3.3).

We denote the displacement in the beam e_j by $u_j = u_j(x, t)$ and the variation of temperature between the actual state and a reference temperature in a thermoelastic beam e_j by $\theta_j = \theta_j(x, t)$, in position x at time t.

Assume that:

- Every thermoelastic beam e_j satisfies the following equations:

$$u_{j,tt} + u_{j,xxxx} - \gamma_j \theta_{j,xx} = 0 \quad \text{in } (0, \ell_j) \times (0, \infty), \tag{3.1}$$

$$\theta_{j,t} - \theta_{j,xx} + \gamma_j u_{j,xxt} = 0 \quad \text{in } (0, \ell_j) \times (0, \infty), \tag{3.2}$$

where γ_j is a positive constant, with initial conditions

$$u_j(x, 0) = u_j^0(x), \quad u_{j,t}(x, 0) = u_j^1(x), \quad \theta_j(x, 0) = \theta_j^0(x). \tag{3.3}$$

- Every elastic beam e_j satisfies the following equation:

$$u_{j,tt} + u_{j,xxxx} = 0 \quad \text{in } (0, \ell_j) \times (0, \infty), \tag{3.4}$$

with initial conditions

$$u_j(x, 0) = u_j^0(x), \quad u_{j,t}(x, 0) = u_j^1(x). \tag{3.5}$$

Denote by J_k^{te} the set of indices of thermoelastic beams adjacent to a_k and J_k^e the set of elastic beams adjacent to a_k. Let $\mathcal{V}_{ext}'$ be the set of external nodes of maximal subgraph of thermoelastic beams. Then, the boundary conditions on the graph $\mathcal{G}$ are described as follows:

$$u_j(a_k, t) = 0, \qquad j \in J_k, \ a_k \in \mathcal{V}_{ext}, \tag{3.6}$$

$$\theta_j(a_k, t) = 0, \qquad j \in J_k, \ a_k \in \mathcal{V}_{ext}', \tag{3.7}$$

$$u_{j,xx}(a_k, t) = 0, \qquad j \in J_k, \ a_k \in \mathcal{V}_{ext}, \tag{3.8}$$

$$u_j(a_k, t) = u_l(a_k, t), \qquad j, l \in J_k, \ a_k \in \mathcal{V}_{int}, \tag{3.9}$$

$$\theta_j(a_k, t) = \theta_l(a_k, t), \qquad j, l \in J_k^{te}, \ a_k \in \mathcal{V}_{int}, \tag{3.10}$$

$$u_{j,xx}(a_k, t) = u_{l,xx}(a_k, t), \qquad j, l \in J_k, \ a_k \in \mathcal{V}_{int}, \tag{3.11}$$

$$\sum_{j \in J_k^{te}} d_{kj} u_{j,x}(a_k, t) = 0, \quad a_k \in \mathcal{V}_{int}, \tag{3.12}$$

$$\sum_{j \in J_k^{te}} d_{kj} \left(u_{j,xxx}(a_k, t) - \gamma_j \theta_{j,x}(a_k, t) \right) + \sum_{j \in J_k^e} d_{kj} u_{j,xxx}(a_k, t) = 0, \quad a_k \in \mathcal{V}_{int}, \tag{3.13}$$

$$\sum_{j \in J_k^{te}} d_{kj} \left(\gamma_j u_{j,xt}(a_k, t) - \theta_{j,x}(a_k, t) \right) = 0, \quad a_k \in \mathcal{V}_{int}. \tag{3.14}$$

Note that (3.9) and (3.10) imply the continuity of $\underline{u}$ and $\underline{\theta}$, and conditions (3.11)–(3.14) are transmission conditions at inner nodes.

We define the total energy of the system by

$$E(t) = \frac{1}{2} \sum_{j=1}^{N} \int_0^{\ell_j} \left(|u_{j,t}|^2 + |u_{j,xx}|^2 \right) dx + \frac{1}{2} \sum_{j \in J^{te}} \int_0^{\ell_j} |\theta_j|^2 dx,$$

where J^{te} is the set of thermoelastic edges; similarly, we will denote by J^e the set of elastic edges.

We prove that

$$E(t) \leq M E(0) e^{-wt} \tag{3.15}$$

for some positive numbers M and w. That is, the energy decays exponentially to 0.

This chapter is organized as follows. Section 1 gathers the functional setting, existence, and uniqueness of solutions. In Sect. 2, we show the exponential stability of the semigroup generated by the system. In Sect. 3, we briefly look at other cases of boundary conditions.

1   Functional Spaces, Existence, and Uniqueness of Solutions

In this section, we consider the well-posedness of the system (3.1)–(3.14). We first rewrite it in a first-order evolution equation.

To start, let us introduce the following spaces:

$$L^2(\mathcal{G}) = \prod_{j=1}^{N} L^2(0, \ell_j); \quad H^k(\mathcal{G}) = \prod_{j=1}^{N} H^k(0, \ell_j), \ k = 2, 4;$$

$$V(\mathcal{G}) = \prod_{j=1}^{N} V_j \text{ where } V_j = L^2(0, \ell_j) \text{ if } j \in J^{te} \text{ and } V_j = \{0\} \text{ if } j \in J^e;$$

$$V^k(\mathcal{G}) = \prod_{j=1}^{N} V_j^k \text{ where } V_j^k = H^k(0, \ell_j) \text{ if } j \in J^{te} \text{ and}$$

$$V_j^k = \{0\} \text{ if } j \in J^e, \ k = 1, 2$$

and

$$F(\mathcal{G}) = \{\underline{f} = (f_1, \dots, f_N) \in H^2(\mathcal{G}) \text{ satisfying } (3.16)\text{–}(3.18)\},$$

$$f_j(a_k) = f_l(a_k), \quad j, l \in J_k, \ a_k \in V_{int}, \tag{3.16}$$

$$f_j(a_k) = 0, \quad j \in J_k, \quad a_k \in \mathcal{V}_{ext}, \tag{3.17}$$

$$\sum_{j \in J_k} d_{kj} \partial_x f_j(a_k) = 0, \quad a_k \in \mathcal{V}_{int}. \tag{3.18}$$

Let the space

$$\mathcal{H} = F(\mathcal{G}) \times L^2(\mathcal{G}) \times V(\mathcal{G}).$$

Lemma 3.1 *The map* $\mathcal{H} \times \mathcal{H} \to \mathbb{C}, (y, \tilde{y}) \mapsto \langle y, \tilde{y} \rangle_\mathcal{H}$ *defined by*

$$\langle y, \tilde{y} \rangle_\mathcal{H} = \sum_{j=1}^N \left(\int_0^{\ell_j} \partial_x^2 f_j(x) \overline{\partial_x^2 \tilde{f}_j}(x) dx + \int_0^{\ell_j} g_j(x) \overline{\tilde{g}_j}(x) dx \right.$$

$$\left. + \int_0^{\ell_j} h_j(x) \overline{\tilde{h}_j}(x) dx \right), \tag{3.19}$$

where $y = (\underline{f}, \underline{g}, \underline{h})$ *and* $\tilde{y} = (\underline{\tilde{f}}, \underline{\tilde{g}}, \underline{\tilde{h}})$ *is an inner product on* $\mathcal{H}$. *Moreover,* $(\mathcal{H}, \langle ., . \rangle_\mathcal{H})$ *is a Hilbert space.*

Proof It suffices to prove that for every $\underline{f} \in F(\mathcal{G})$ the assumption

$$\sum_{j=1}^N \left(\int_0^{\ell_j} \partial_x^2 f_j(x) \overline{\partial_x^2 f_j(x)} dx \right) = 0 \tag{3.20}$$

implies $\underline{f} = 0$.

Let $\underline{f}$ in $F(\mathcal{G})$ satisfying (3.20).

We use a matrix method [98]. First of all, we recall some definitions and notations already presented in the preliminary chapter. The incidence matrix $D = (d_{kj})_{p \times N}$ and the adjacency matrix $E = (e_{ik})_{p \times p}$ are defined by,

$$d_{kj} = \begin{cases} 1 & \text{if } \pi_j(\ell_j) = a_k \\ -1 & \text{if } \pi_j(0) = a_k \\ 0 & \text{otherwise} \end{cases} \quad \text{and} \quad e_{ik} = \begin{cases} 1 & \text{if } a_i \text{ and } a_k \text{ are adjacent} \\ 0 & \text{otherwise}. \end{cases}$$

For two matrices $A = (a_{jk})$ and $B = (b_{jk})$ of the same size,

$$A * B = (a_{jk} b_{jk}),$$

and for any function $Q : \mathbb{R} \longrightarrow \mathbb{R}$, we define the matrix $Q(A) = (q_{ik})_{p \times p}$ by

$$q_{ik} = \begin{cases} Q(a_{ik}) & \text{if } e_{ik} = 1, \\ 0 & \text{otherwise}. \end{cases}$$

In particular, we write $A^{(r)} = Q(A)$ if $Q(x) = x^r$. Furthermore, the matrix $L = (\ell_{ik})_{p \times p}$, is defined by

$$\ell_{ik} = \begin{cases} \ell_{s(i,k)} & \text{if } e_{ik} = 1, \\ 0 & \text{otherwise,} \end{cases}$$

where $s(i, k) = s(k, i)$ is the index of the edge connecting a_i and a_k.

Now, we introduce the following definition used in [98] (see also [1]):

To the function $\underline{f}$ is associated the matrix function F defined by

$$F : [0, 1] \longrightarrow \mathbb{C}^{p \times p}, x \longmapsto F(x) = (f_{jk}(x))_{p \times p},$$

with

$$f_{jk}(x) = e_{jk} f_{s(j,k)} \left[\ell_{s(j,k)} \left(\frac{1 + d_{js(j,k)}}{2} - x d_{js(j,k)} \right) \right].$$

The continuity condition of $\underline{f}$ at the interior nodes and Dirichlet conditions at the external nodes can be expressed in this manner:

There exists $\varphi = \begin{pmatrix} \varphi_1 \\ \vdots \\ \varphi_p \end{pmatrix} \in \mathbb{C}^p$ such that $\varphi_k = 0$ when a_k is an external node,

and

$$F(0) = (\varphi e^T) * E_{int}, \tag{3.21}$$

where $e = \begin{pmatrix} 1 \\ \vdots \\ 1 \end{pmatrix} \in \mathbb{R}^n$ and $E_{int} = (e'_{jk})_{p \times p}$ is the matrix defined by

$$e'_{jk} = \begin{cases} 1 & \text{if } a_j \text{ and } a_k \text{ are adjacent and } |J_k| > 1, \\ 0 & \text{otherwise.} \end{cases}$$

Recall that the matrix $L^{(-1)} = (L_{ik})_{p \times p}$ is defined as follows:

$$L_{ik} = \begin{cases} \frac{1}{\ell_{s(i,k)}} & \text{if } e_{ik} = 1, \\ 0 & \text{otherwise,} \end{cases}$$

then the condition (3.18) applied to $\underline{f}$ is expressed as follows:

$$(L^{(-1)} * F'(0) * E_{int})e = 0. \tag{3.22}$$

Finally, we have

$$F(1 - x) = F(x)^T. \tag{3.23}$$

From (3.20), we deduce the following matrix-differential equation:

$$F(x) = xL * C + D, \tag{3.24}$$

where

$$C = L^{(-1)} * F'(0) \text{ and } D = F(0).$$

By taking $x = 1$ in (3.24) and using (3.23), we obtain

$$F(1) = D^T = L * C + D$$

which implies that

$$D^T - D = L * C,$$

then, using (3.22), we get

$$\left(L^{(-1)} * (D^T - D) * E_{int}\right) e = 0. \tag{3.25}$$

We recall the following elementary rules for a matrix $M \in \mathbb{C}^{p \times p}$ (see [22]):

$$(M * D^T)e = M\varphi, \quad (M * D)e = diag(Me)\varphi. \tag{3.26}$$

Then, (3.25) implies that

$$K\varphi = 0,$$

where

$$K = L^{(-1)} * E_{int} - diag\left(\left(L^{(-1)} * E_{int}\right)e\right).$$

Because $\mathcal{V}_{ext} \neq \emptyset$, the matrix obtained from $K * E_{int}^T * E_{int}$ by removing rows and columns that are zero is invertible [1]. By taking into account that $\varphi_j = 0$ when a_j is an exterior vertex, we deduce that $\varphi = 0$. Then $D = 0$ and $C = 0$. Hence, $\underline{f = 0}$.   □

Now, define the linear operator $\mathcal{A} : \mathcal{D}(\mathcal{A}) \subseteq \mathcal{H} \to \mathcal{H}$,

$$\mathcal{A} = \begin{pmatrix} 0 & A_1^0 & 0 \\ -A_1^4 & 0 & A_\gamma^2 \\ 0 & -A_\gamma^2 & A_1^2 \end{pmatrix}, \tag{3.27}$$

where $A_\gamma^k = diag(\gamma_1 \partial_x^k, \ldots, \gamma_N \partial_x^k)$, $k \in \mathbb{N}$, and $\partial_x^0 = I$, and whose domain is given by

$$\mathcal{D}(\mathcal{A}) = \left\{ (\underline{u}, \underline{v}, \underline{\theta}) \in \left(F(\mathcal{G}) \cap H^4(\mathcal{G}) \right) \times F(\mathcal{G}) \times V^2(\mathcal{G}) \right.$$

satisfying (3.28)–(3.33) below$\}$

$$\partial_x^2 u_j(a_k) = \partial_x^2 u_l(a_k), \qquad j, l \in J_k, \ a_k \in \mathcal{V}_{int}, \tag{3.28}$$

$$\partial_x^2 u_j(a_k) = 0, \qquad\qquad j \in J_k, \ a_k \in \mathcal{V}_{ext}, \tag{3.29}$$

$$\theta_j(a_k) = \theta_l(a_k), \qquad j, l \in J_k^{te}, \ a_k \in \mathcal{V}_{int}, \tag{3.30}$$

$$\theta_j(a_k) = 0, \qquad\qquad j \in J_k, \ a_k \in \mathcal{V}'_{ext}, \tag{3.31}$$

$$\sum_{j \in J_k} d_{kj} \left(\partial_x^3 u_j(a_k) - \gamma_j \partial_x \theta_j(a_k) \right) = 0, \ \ a_k \in \mathcal{V}_{int}, \tag{3.32}$$

$$\sum_{j \in J_k} d_{kj} \left(\gamma_j \partial_x v_j(a_k) - \partial_x \theta_j(a_k) \right) = 0, \ \ a_k \in \mathcal{V}_{int}, \tag{3.33}$$

with $\gamma_j = 0$ if $j \in J^e$.

For $y = (\underline{u}, \underline{v}, \underline{\theta}) \in \mathcal{D}(\mathcal{A})$, the components of $\mathcal{A}y$ are

$$\begin{cases} v_j, & j = 1, \ldots, N, \\ -\partial_x^4 u_j + \gamma_j \partial_x^2 \theta_j, & j = 1, \ldots, N, \\ -\gamma_j \partial_x^2 v_j + \partial_x^2 \theta_j, & j = 1, \ldots, N. \end{cases}$$

So that, the initial boundary value problem (3.1)–(3.14) can be written as an evolutionary equation in $\mathcal{H}$

$$\begin{cases} \dfrac{d}{dt} y = \mathcal{A}y, \\ y(0) = y^0, \end{cases} \tag{3.34}$$

where $y = (\underline{u}, \underline{u}_t, \underline{\theta})$ and $y^0 = (\underline{u}^0, \underline{u}^1, \underline{\theta}^0)$.

Lemma 3.2 *Let $\mathcal{H}$ and $\mathcal{A}$ be defined as before. Then*

(i) The operator $\mathcal{A}$ is dissipative.

(ii) $1 \in \rho(\mathcal{A})$: the resolvent set of $\mathcal{A}$, and $(\mathcal{I} - \mathcal{A})^{-1}$ is compact.

Proof (i) Let $y = (\underline{u}, \underline{v}, \underline{w}) \in \mathcal{D}(\mathcal{A})$. Using the definition of the inner product (3.19), we have

$$Re(\langle \mathcal{A}y, y \rangle_{\mathcal{H}}) = Re \sum_{j=1}^{N} \left(\int_0^{\ell_j} \partial_x^2 v_j \partial_x^2 \overline{u_j} dx + \int_0^{\ell_j} (-\partial_x^4 u_j + \gamma_j \partial_x^2 \theta_j) \overline{v_j} dx \right.$$

$$\left. + \int_0^{\ell_j} (-\gamma_j \partial_x^2 v_j + \partial_x^2 \theta_j) \overline{\theta_j} dx \right).$$

Performing some integrations by parts,

$$Re(\langle \mathcal{A}y, y \rangle_{\mathcal{H}}) = Re \left(\sum_{j=1}^{N} \left(\partial_x^2 u_j \partial_x \overline{v_j} \Big|_{x=0}^{x=\ell_j} + (-\partial_x^3 u_j + \gamma_j \partial_x \theta_j) \overline{v_j} \Big|_{x=0}^{x=\ell_j} \right. \right.$$

$$\left. \left. - \gamma_j \, \theta_j \partial_x \overline{v_j} \Big|_{x=0}^{x=\ell_j} + \theta_j \partial_x \overline{\theta_j} \Big|_{x=0}^{x=\ell_j} - \left\| \partial_x \theta_j \right\|^2 \right) \right). \tag{3.35}$$

Using boundary conditions, it yields

$$Re(\langle \mathcal{A}y, y \rangle_{\mathcal{H}}) = Re \left(\sum_{k=1}^{n} \left[\sum_{j \in J_k} \left(\overline{v_j}(a_k)(-\partial_x^3 u_j(a_k) + \gamma_j \partial_x \theta_j(a_k)) \right) \right. \right.$$

$$\left. \left. + \sum_{j \in J_k^{te}} \left(\overline{\theta_j}(a_k)(-\gamma_j \partial_x v_j(a_k) + \partial_x \theta_j(a_k)) \right) \right] \right)$$

$$- \sum_{j=1}^{N} \left\| \partial_x \theta_j \right\|^2$$

$$= - \sum_{j=1}^{N} \left\| \partial_x \theta_j \right\|^2 \leq 0, \tag{3.36}$$

and this proves the dissipativeness of the operator $\mathcal{A}$ in $\mathcal{H}$.

Next, we shall prove that $1 \in \rho(\mathcal{A})$. Let $z = (\underline{f}, \underline{g}, \underline{h}) \in \mathcal{H}$, we look for $y = (\underline{u}, \underline{v}, \underline{\theta}) \in \mathcal{D}(\mathcal{A})$ such that

$$(\mathcal{I} - \mathcal{A})y = z \tag{3.37}$$

i.e.,

$$u_j - v_j = f_j, \quad j = 1, \ldots, N, \tag{3.38}$$

$$v_j + \partial_x^4 u_j - \gamma_j \partial_x^2 \theta_j = g_j, \quad j = 1, \ldots, N, \tag{3.39}$$

$$\theta_j - \partial_x^2 \theta_j + \gamma_j \partial_x^2 v_j = h_j, \quad j = 1, \ldots, N. \tag{3.40}$$

Substituting (3.38) into (3.39) and (3.40), we obtain

$$u_j + \partial_x^4 u_j - \gamma_j \partial_x^2 \theta_j = g_j + f_j, \quad j = 1, \ldots, N, \tag{3.41}$$

$$\theta_j - \partial_x^2 \theta_j + \gamma_j \partial_x^2 u_j = h_j + \gamma_j \partial_x^2 f_j, \quad j = 1, \ldots, N. \tag{3.42}$$

For the sequel, we need the following space:

$$\mathcal{F} = \left\{ (\underline{w}, \underline{\varphi}) \in H^2(\mathcal{G}) \times V^1(\mathcal{G}) \text{ satisfying (3.43) below} \right\}$$

$$\begin{cases} w_j(a_k) = w_l(a_k), & j, l \in J_k, \ a_k \in \mathcal{V}_{int}, \\ \varphi_j(a_k) = \varphi_l(a_k), & j, l \in J_k^{te}, \ a_k \in \mathcal{V}_{int}, \\ w_j(a_k) = 0, & j \in J_k, \ a_k \in \mathcal{V}_{ext}, \\ \varphi_j(a_k) = 0, & j \in J_k, \ a_k \in \mathcal{V}_{ext}', \\ \sum_{j \in J_k} d_{kj} \partial_x w_j(a_k) = 0, & a_k \in \mathcal{V}_{int}. \end{cases} \tag{3.43}$$

Equipped with the inner product

$$\left\langle (\underline{w}, \underline{\varphi}), (\underline{\tilde{w}}, \underline{\tilde{\varphi}}) \right\rangle_{\mathcal{F}} = \sum_{j=1}^{N} \int_0^{\ell_j} \partial_x^2 w_j \partial_x^2 \overline{\tilde{w}}_j dx + \sum_{j=1}^{N} \int_0^{\ell_j} \partial_x \varphi_j \partial_x \overline{\tilde{\varphi}}_j dx,$$

$\mathcal{F}$ is a Hilbert space.

For $(\underline{w}, \underline{\varphi})$ in $\mathcal{F}$. Taking the inner product, in $L^2(0, \ell_j)$, of (3.41) by w_j and (3.42) by φ_j, $j = 1, \ldots, N$, and integrating by parts, we obtain, respectively

$$\int_0^{\ell_j} u_j \overline{w}_j dx + \int_0^{\ell_j} \partial_x^2 u_j \partial_x^2 \overline{w}_j dx + (\partial_x^3 u_j - \gamma_j \partial_x \theta_j) \overline{w}_j \Big|_{x=0}^{x=\ell_j}$$

$$- \partial_x^2 u_j \partial_x \overline{w}_j \Big|_{x=0}^{x=\ell_j} + \gamma_j \int_0^{\ell_j} \partial_x \theta_j \partial_x \overline{w}_j dx = \int_0^{\ell_j} (g_j + f_j) \overline{w}_j dx$$

and

$$\int_0^{\ell_j} \theta_j \overline{\varphi}_j dx + (-\partial_x \theta_j + \gamma_j \partial_x u_j) \overline{\varphi}_j \Big|_{x=0}^{x=\ell_j} + \int_0^{\ell_j} \partial_x \theta_j \partial_x \overline{\varphi}_j dx$$

$$- \gamma_j \int_0^{\ell_j} \partial_x u_j \partial_x \overline{\varphi}_j dx = \int_0^{\ell_j} (h_j + \gamma_j \partial_x^2 f_j) \overline{\varphi}_j dx$$

for $j = 1, \ldots, N$. Now summing up over $j \in \{1, \ldots, N\}$, we find by taking into account the boundary conditions (3.16)–(3.18), (3.28)–(3.33), and (3.43),

$$a((\underline{u}, \underline{\theta}), (\underline{w}, \underline{\varphi})) = g(\underline{w}, \underline{\varphi}), \tag{3.44}$$

where

$$a((\underline{u},\underline{\theta}),(\underline{w},\underline{\varphi})) = \sum_{j=1}^{N}\int_{0}^{\ell_j} u_j\overline{w_j}dx + \sum_{j=1}^{N}\int_{0}^{\ell_j}\partial_x^2 u_j\partial_x^2\overline{w_j}dx + \sum_{j=1}^{N}\int_{0}^{\ell_j}\theta_j\overline{\varphi_j}dx$$

$$+ \sum_{j=1}^{N}\int_{0}^{\ell_j}\partial_x\theta_j\partial_x\overline{\varphi_j}dx + \gamma_j\left(\int_{0}^{\ell_j}(\partial_x\theta_j\partial_x\overline{w_j} - \partial_x u_j\partial_x\overline{\varphi_j})dx\right)$$

and

$$g(\underline{w},\underline{\varphi}) = \sum_{j=1}^{N}\int_{0}^{\ell_j}(g_j + f_j)\overline{w_j}dx + \sum_{j=1}^{N}\int_{0}^{\ell_j}(h_j + \gamma_j\partial_x^2 f_j)\overline{\varphi_j}dx$$

$$+ \sum_{a_k\in\mathcal{V}_{int}}\sum_{j\in J_k}\gamma_j\partial_x f_j(a_k))\overline{\varphi_j}(a_k).$$

a is a continuous sesquilinear form on $\mathcal{F}\times\mathcal{F}$ and g is a continuous anti-linear form on $\mathcal{F}$. Moreover, for every $(\underline{w},\underline{\varphi})\in\mathcal{F}$

$$\left|a((\underline{u},\underline{\theta}),(\underline{w},\underline{\varphi}))\right| \geq \left\|(\underline{w},\underline{\varphi})\right\|_{\mathcal{F}}^{2}.$$

It follows, using the Lax–Milgram's lemma, that (3.44) has a unique solution $(\underline{u},\underline{\theta})$ in $\mathcal{F}$.

If we consider respectively $(\underline{w},\underline{\varphi})$ in $\{0\}\times\prod_{j=1}^{N}\mathcal{D}(0,\ell_j)$ and (w,φ) in $\prod_{j=1}^{N}\mathcal{D}(0,\ell_j)\times\{0\}$ then, $(\underline{u},\underline{\theta})$ belongs to the space $H^4(\mathcal{G})\times H^2(\mathcal{G})$ and satisfies

$$u_j + \partial_x^4 u_j - \gamma_j\partial_x^2\theta_j = g_j + f_j, \quad j = 1,\ldots,N,$$
$$\theta_j - \partial_x^2\theta_j + \gamma_j\partial_x^2 u_j = h_j + \gamma_j\partial_x^2 f_j, \quad j = 1,\ldots,N.$$

By some integrations by parts, we deduce that the solution $(\underline{u},\underline{\theta})$ satisfies the conditions

$$\begin{cases} \partial_x^2 u_j(a_k) = \partial_x^2 u_l(a_k), & j,l\in J_k, \ a_k\in\mathcal{V}_{int}, \\ \partial_x^2 u_j(a_k) = 0, & j\in J_k, \ a_k\in\mathcal{V}_{ext}, \\ \sum_{j\in J_k} d_{kj}\left(\partial_x^3 u_j(a_k) - \gamma_j\partial_x\theta_j(a_k)\right) = 0, & a_k\in\mathcal{V}_{int}, \\ \sum_{j\in J_k} d_{kj}\left(-\gamma_j\partial_x u_j(a_k) + \partial_x\theta_j(a_k)\right) = \sum_{j\in J_k} d_{kj}\gamma_j\partial_x f_j(a_k), & a_k\in\mathcal{V}_{int}. \end{cases}$$

Returning back to the Lax–Milgram's lemma, $y = (\underline{u}, \underline{v}, \underline{\theta})$ verifies

$$\|y\|_{\mathcal{H}}^2 \le c \, \|z\|_{\mathcal{H}}^2 \,,$$

where c is a positive constant independent of y. All of that prove that $y = (\underline{u}, \underline{v}, \underline{\theta}) \in \mathcal{D}(\mathcal{A})$ and $(\mathcal{I} - \mathcal{A})^{-1} \in \mathcal{L}(\mathcal{H})$, that is, $1 \in \rho(A)$.

Moreover, one can deduce, by the Sobolev embedding theorem, that $(\mathcal{I} - \mathcal{A})^{-1}$ is a compact operator. The proof is complete.   □

It follows from the Lumer–Phillips theorem (Theorem 1.15):

Corollary 3.3 *The operator $\mathcal{A}$ is the infinitesimal generator of a C_0-semigroup of contraction $(S(t))_{t \ge 0}$ on the Hilbert space $\mathcal{H}$.*
Hence, for any $y^0 \in \mathcal{H}$, the Cauchy problem (3.34) has a unique solution

$$y \in C([0, +\infty), \mathcal{H}).$$

Furthermore, if $y^0 \in \mathcal{D}(\mathcal{A})$, then

$$y \in C([0, +\infty), \mathcal{D}(\mathcal{A})) \cap C^1([0, +\infty), \mathcal{H}).$$

2   Exponential Decay

The aim of this section is to prove the exponential decay rate of the energy $E(t)$ of the whole system, i.e., there exist two positive constants M and w verifying

$$E(t) \le M E(0) e^{-wt} \tag{3.45}$$

or equivalently, there are two positive constants M and w such that

$$\|S(t)\| \le M e^{-wt}, \quad \forall t \ge 0.$$

To prove such property we use Theorem 1.25. We will first verify the condition (1.9) in Theorem 1.25.

Lemma 3.4 *Let $\mathcal{A}$ be the operator given by (3.27), then the assumption (1.9) holds, that is $i\mathbb{R} \subset \rho(\mathcal{A})$.*

Proof Suppose that (1.9) is not true. Then, there is a real number $\beta \in \mathbb{R}$ such that $\lambda := i\beta$ is in $\sigma(\mathcal{A})$. Since $(\mathcal{I} - \mathcal{A})^{-1}$ is compact, λ must be an eigenvalue of $\mathcal{A}$, then there is a vector $y = (\underline{u}, \underline{v}, \underline{\theta}) \in \mathcal{D}(\mathcal{A})$, $y \ne 0$ such that $\mathcal{A}y = i\beta y$.

Already, we have

$$Re(\langle \mathcal{A}y, y \rangle) = Re(\langle \lambda y, y \rangle) = 0,$$

which leads to

$$-\sum_{j=1}^{N} \left\| \partial_x \theta_j \right\|^2 = 0.$$

Then, using the continuity condition of $\underline{\theta}$ at inner nodes, the Dirichlet condition of $\underline{\theta}$ in $\mathcal{V}'_{ext}$, and the condition that every maximal subgraph of thermoelastic edges is not a circuit, we deduce that $\theta_j = 0$ for $j = 1, \ldots, N$. Thus, $(\underline{u}, \underline{v})$ satisfies

$$\begin{cases} v_j = \lambda u_j, & j = 1, \ldots, N, \\ -\partial_x^4 u_j = \lambda^2 u_j, & j = 1, \ldots, N, \\ -\gamma_j \partial_x^2 v_j = 0, & j = 1, \ldots, N. \end{cases} \tag{3.46}$$

If $\lambda \neq 0$ and e_j is thermoelastic, then by the third and first equations in (3.46), $\partial_x^2 u_j = 0$ and $v_j = 0$ in $L^2(0, \ell_j)$. Furthermore, $\partial_x^k u_j(a_k)$ for $k = 0, \ldots, 3$ and if a_k is an end of e_j.

If $\lambda \neq 0$ and e_j is the only elastic edge attached to a thermoelastic edge at the end a_k, then u_j satisfies the Cauchy problem:

$$\partial_x^4 u_j + \lambda^2 u_j = 0 \text{ and } \partial_x^k u_j(a_k) = 0, \ k = 0, \ldots, 3.$$

So that $\partial_x^k u_j, k = 0, \ldots, 3$ are zero in $L^2(0, \ell_j)$ and at both ends of e_j. We iterate this procedure in each maximal subgraph of elastic edges of $\mathcal{G}$; we then conclude that $u_j = 0$ in $H^2(0, \ell_j)$ and $v_j = 0$ in $L^2(0, \ell_j)$.

Now, suppose that $\lambda = 0$. Then,

$$\begin{cases} v_j = 0, & j = 1, \ldots, N, \\ -\partial_x^4 u_j = 0, & j = 1, \ldots, N, \\ -\gamma_j \partial_x^2 v_j = 0, & j = 1, \ldots, N. \end{cases} \tag{3.47}$$

Multiplying the second equation in the above system by u_j and then summing over j, we obtain, by taking into account conditions (3.18) and (3.32)

$$\sum_{j=1}^{N} \left\| \partial_x^2 u_j \right\|^2 = 0$$

which implies that $\underline{u} = 0$ in $H^2(\mathcal{G})$. $\qquad\square$

Next we prove that $S(t)_{t \geq 0}$ satisfies (1.10) in Theorem 1.25.

Lemma 3.5 *Let $\mathcal{A}$ be the operator given by (3.27), then condition (1.10) holds for the semigroup $(S(t))_{t \geq 0}$.*

Proof Suppose that (1.10) is not true, then there exists a sequence (β_n) of real numbers, with $\beta_n \longrightarrow \infty$ ($\beta_n > 0$ without loss of generality) and a sequence of vectors $(y_n) = (\underline{u}_n, \underline{v}_n, \underline{\theta}_n)$ in $\mathcal{D}(\mathcal{A})$ with $\|y_n\|_{\mathcal{H}} = 1$, such that

$$\|(\mathrm{i}\beta_n - \mathcal{A})y_n\|_{\mathcal{H}} \longrightarrow 0.$$

Writing this condition term by term we get,

$$\mathrm{i}\beta_n u_{j,n} - v_{j,n} = f_{j,n} \longrightarrow 0, \quad \text{in } H^2(0, \ell_j), \qquad (3.48)$$

$$\mathrm{i}\beta_n v_{j,n} + \partial_x^4 u_{j,n} - \gamma_j \partial_x^2 \theta_{j,n} = g_{j,n} \longrightarrow 0, \quad \text{in } L^2(0, \ell_j), \qquad (3.49)$$

$$\mathrm{i}\beta_n \theta_{j,n} - \partial_x^2 \theta_{j,n} + \gamma_j \partial_x^2 v_{j,n} = h_{j,n} \longrightarrow 0, \quad \text{in } L^2(0, \ell_j), \qquad (3.50)$$

for j in $\{1, \ldots, N\}$.

Step 1 Since $Re(\langle(\mathrm{i}\beta_n - \mathcal{A})y_n, y_n\rangle_{\mathcal{H}}) = \sum_{j=1}^{N} \|\partial_x \theta_{j,n}\|$, we obtain

$$\|\partial_x \theta_{j,n}\| \longrightarrow 0, \ \text{ for } j = 1, \ldots, N$$

which implies

$$\theta_{j,n} \longrightarrow 0, \ \text{ in } L^2(0, \ell_j)$$

for $j = 1, \ldots, N$, by (3.30), (3.31) and the fact that every maximal subgraph of thermoelastic edges is not a circuit.

Step 2 We will prove that $\partial_x^2 u_{j,n} \longrightarrow 0$ and $v_{j,n} \longrightarrow 0$ in $L^2(0, \ell_j)$ for j in J^{te}.

Let j in J^{te}. From (3.48) we deduce that $\left\|\frac{\partial_x^2 v_{j,n}}{\beta_n}\right\|$ is bounded. Then it is easy to deduce, respectively, from (3.50) and (3.49), the boundedness of $\left\|\frac{\partial_x^2 \theta_{j,n}}{\beta_n}\right\|$ and $\left\|\frac{\partial_x^4 u_{j,n}}{\beta_n}\right\|$.

Using (3.48) again to replace $\partial_x^2 v_{j,n}$ by $\beta_n \partial_x^2 u_{j,n}$ in (3.50), multiplying the new equation by $\frac{\partial_x^2 u_{j,n}}{\beta_n}$, we obtain, using the fact that $\theta_{j,n} \longrightarrow 0$ and that $\partial_x^2 u_{j,n}$ is bounded in $L^2(0, \ell_j)$,

$$\gamma_j \left\|\partial_x^2 u_{j,n}\right\|^2 - \frac{1}{\beta_n}\left\langle \partial_x^2 \theta_{j,n}, \gamma_j \partial_x^2 u_{j,n}\right\rangle \longrightarrow 0 \ \text{ in } L^2(0, \ell_j).$$

Integrating by parts, we obtain,

$$\gamma_j \left\|\partial_x^2 u_{j,n}\right\|^2 - \frac{\gamma_j}{\beta_n} \partial_x \theta_{j,n} \partial_x^2 \overline{u_{j,n}}\Big|_{x=0}^{x=\ell_j} + \frac{\gamma_j}{\beta_n}\left\langle \partial_x \theta_{j,n}, \partial_x^3 u_{j,n}\right\rangle \longrightarrow 0. \qquad (3.51)$$

We want prove that the second and third terms in the left hand side of the above expression tend to 0 as n tends to ∞. To do this we will apply Theorem 1.33 more than one time.

Inequality (1.14), applied to $w = \partial_x^2 u_{j,n}$ gives

$$\frac{\left\|\partial_x^3 u_{j,n}\right\|}{\beta_n^{1/2}} \leq C_3 \left\|\frac{\partial_x^4 u_{j,n}}{\beta_n}\right\|^{1/2} \left\|\partial_x^2 u_{j,n}\right\|^{1/2} + C_4 \frac{\left\|\partial_x^2 u_{j,n}\right\|}{\beta_n^{1/2}}. \tag{3.52}$$

It follows that $\frac{\left\|\partial_x^3 u_{j,n}\right\|}{\beta_n^{1/2}}$ is bounded and then the third term in (3.51) tends to zero, since $\partial_x \theta_{j,n} \longrightarrow 0$ in $L^2(0, \ell_j)$.

Inequality (1.13) applied, respectively, to $w = \partial_x^2 u_{j,n}$ and $w = \partial_x \theta_{j,n}$ gives, respectively,

$$\frac{\left\|\partial_x^2 u_{j,n}\right\|_\infty}{\beta_n^{1/4}} \leq C_1 \left\|\partial_x^2 u_{j,n}\right\|^{1/2} \left\|\frac{\partial_x^3 u_{j,n}}{\beta_n^{1/2}}\right\|^{1/2} + C_2 \frac{\left\|\partial_x^2 u_{j,n}\right\|}{\beta_n^{1/4}}$$

and

$$\frac{\left\|\partial_x \theta_{j,n}\right\|_\infty}{\beta_n^{1/2}} \leq C_1 \left\|\partial_x \theta_{j,n}\right\|^{1/2} \left\|\frac{\partial_x^2 \theta_{j,n}}{\beta_n}\right\|^{1/2} + C_2 \frac{\left\|\partial_x \theta_{j,n}\right\|}{\beta_n^{1/2}}$$

which imply that $\frac{\left\|\partial_x^2 u_{j,n}\right\|_\infty}{\beta_n^{1/4}}$ is bounded and $\frac{\left\|\partial_x \theta_{j,n}\right\|_\infty}{\beta_n^{1/2}}$ tends to zero. Hence, the second term in (3.51) converges to zero. Thus, (3.51) is reduced to

$$\left\|\partial_x^2 u_{j,n}\right\|^2 \longrightarrow 0. \tag{3.53}$$

Going back to (3.48) and (3.50), we get, respectively,

$$\frac{\partial_x^2 v_{j,n}}{\beta_n} \longrightarrow 0 \text{ in } L^2(0, \ell_j)$$

and

$$\frac{\partial_x^2 \theta_{j,n}}{\beta_n} \longrightarrow 0 \text{ in } L^2(0, \ell_j).$$

Dividing (3.49) by β_n, multiplying the result by $v_{j,n}$ and integrating by parts, we obtain

$$\mathbf{i} \left\|v_{j,n}\right\|^2 + \frac{1}{\beta_n} \left. \partial_x^3 u_{j,n} \overline{v_{j,n}} \right|_{x=0}^{x=\ell_j} - \frac{1}{\beta_n} \left\langle \partial_x^3 u_{j,n}, \partial_x v_{j,n} \right\rangle \longrightarrow 0. \tag{3.54}$$

Rewriting again (1.13) with $w = \partial_x^3 u_{j,n}$ and $w = v_{j,n}$, respectively, and (1.14) with $w = \partial_x v_{j,n}$, dividing the three inequalities, respectively, by $\beta_n^{3/4}$, $\beta_n^{1/4}$, and $\beta_n^{1/2}$, we deduce, using (3.52) and the boundedness of $\frac{\partial_x^4 u_{j,n}}{\beta_n}$, that the second and the third term in (3.54) converge to zero. It follows that

$$v_{j,n} \longrightarrow 0 \quad \text{in } L^2(0, \ell_j).$$

Step 3 We will prove that $\partial_x^2 u_{j,n} \longrightarrow 0$ and $v_{j,n} \longrightarrow 0$ in $L^2(0, \ell_j)$ for j in J^e.

Let j in $\{1, \ldots, N\}$. Combining (3.48) and (3.49), we obtain

$$- \beta_n^2 u_{j,n} + \partial_x^4 u_{j,n} - \gamma_j \partial_x^2 \theta_{j,n} = g_{j,n} + \mathbf{i}\beta_n f_{j,n}. \tag{3.55}$$

Let q be a function in $C^2([0, \ell_j], \mathbb{C})$ independent of n such that $\partial_x^2 q = 0$. Taking the inner product in $L^2(0, \ell_j)$ of (3.55) with $q \partial_x u_{j,n}$, and integrating by parts, we obtain

$$- \frac{1}{2}\beta_n^2 \left| u_{j,n}(x) \right|^2 q(x) \Big|_{x=0}^{x=\ell_j} - \frac{1}{2} \left| \partial_x^2 u_{j,n}(x) \right|^2 q(x) \Big|_{x=0}^{x=\ell_j}$$

$$+ Re \left(\partial_x^3 u_{j,n}(x) q(x) \partial_x \overline{u_{j,n}}(x) \Big|_{x=0}^{x=\ell_j} \right) - Re \left(i\beta_n f_{j,n}(x) q(x) \overline{u_{j,n}}(x) \Big|_{x=0}^{x=\ell_j} \right)$$

$$- \gamma_j Re \left(\partial_x \theta_{j,n}(x) q(x) \partial_x \overline{u_{j,n}}(x) \Big|_{x=0}^{x=\ell_j} \right) + \frac{1}{2} \int_0^{\ell_j} \beta_n^2 \left| u_{j,n} \right|^2 \partial_x q \, dx$$

$$+ \frac{3}{2} \int_0^{\ell_j} \left| \partial_x^2 u_{j,n} \right|^2 \partial_x q \, dx = \int_0^{\ell_j} g_{j,n} q \partial_x \overline{u_{j,n}} \, dx - \int_0^{\ell_j} i\beta_n \overline{u_{j,n}} \partial_x (q f_{j,n}) \, dx$$

$$- \int_0^{\ell_j} \partial_x \theta_{j,n} \partial_x (q \partial_x \overline{u_{j,n}}) \, dx + Re \left(\partial_x^2 u_{j,n}(x) \partial_x \overline{u_{j,n}}(x) \partial_x q(x) \Big|_{x=0}^{x=\ell_j} \right). \tag{3.56}$$

It is easy to verify that the first three terms in the right hand side of the above equation converge to zero.

In the case of thermoelastic edge e_j, the last term in (3.56) converges to zero since $\left\| \frac{\partial_x^2 u_{j,n}}{\beta_n^{1/4}} \right\|_\infty$ and $\left\| \beta_n^{1/4} \partial_x \overline{u_{j,n}} \right\|_\infty$ converge to zero by (1.13). Thus, equation (3.56) is reduced to

$$- \frac{1}{2}\beta_n^2 \left| u_{j,n}(x) \right|^2 q(x) \Big|_{x=0}^{x=\ell_j} - \frac{1}{2} \left| \partial_x^2 u_{j,n}(x) \right|^2 q(x) \Big|_{x=0}^{x=\ell_j}$$

$$+ Re \left(\partial_x^3 u_{j,n}(x) q(x) \partial_x \overline{u_{j,n}}(x) \Big|_{x=0}^{x=\ell_j} \right) - Re \left(\mathbf{i}\beta_n f_{j,n}(x) q(x) \overline{u_{j,n}}(x) \Big|_{x=0}^{x=\ell_j} \right)$$

$$-\gamma_j \, Re\left(\partial_x \theta_{j,n}(x) q(x) \partial_x \overline{u_{j,n}}(x)\big|_{x=0}^{x=\ell_j}\right) + \frac{1}{2}\int_0^{\ell_j} \beta_n^2 \left|u_{j,n}\right|^2 \partial_x q \, dx$$

$$+ \frac{3}{2}\int_0^{\ell_j} \left|\partial_x^2 u_{j,n}\right|^2 \partial_x q \, dx \longrightarrow 0. \tag{3.57}$$

In particular, if a_k is an end of e_j then by taking $q(x) = x$ or $q(x) = \ell_j - x$ we have using results of step 1 and step 2,

$$-\frac{1}{2}\beta_n^2 \left|u_{j,n}(a_k)\right|^2 - \frac{1}{2}\left|\partial_x^2 u_{j,n}(a_k)\right|^2 + Re\left(\partial_x^3 u_{j,n}(a_k)\partial_x \overline{u_{j,n}}(a_k)\right)$$

$$- Re\left(i\beta_n f_{j,n}(a_k)\overline{u_{j,n}}(a_k)\right) - \gamma_j \, Re\left(\partial_x \theta_{j,n}(a_k)\partial_x \overline{u_{j,n}}(a_k)\right) \longrightarrow 0. \tag{3.58}$$

Now, we show that for every interior end a_k of a thermoelastic edge e_j, $\beta_n^{1/2}\partial_x u_{j,n}(a_k)$, $\frac{\partial_x^3 u_{j,n}(a_k)}{\beta_n^{1/2}}$, $\partial_x^2 u_{j,n}(a_k)$, and $\beta_n^j u_{j,n}(a_k)$ tend to zero.

Let j be in $\{1, \ldots, N\}$ and take the inner product of (3.55) with $\frac{1}{\beta_n^{1/2}}e^{-\beta_n^{1/2}(\ell_j - x)}$, yields

$$\int_0^{\ell_j} \beta_n^{3/2}e^{-\beta_n^{1/2}(\ell_j - x)}u_{j,n}(x)dx - \frac{1}{\beta_n^{1/2}}\int_0^{\ell_j} e^{-\beta_n^{1/2}(\ell_j - x)}\partial_x^4 u_{j,n}(x)dx$$

$$= \frac{-1}{\beta_n^{1/2}}\int_0^{\ell_j} \left(g_{j,n} + i\beta_n f_{j,n}\right)e^{-\beta_n^{1/2}(\ell_j - x)}dx$$

$$+ \frac{\gamma_j}{\beta_n^{1/2}}\partial_x \theta_{j,n}(x)e^{-\beta_n^{1/2}(\ell_j - x)}\Big|_{x=0}^{x=\ell_j} - \gamma_j \int_0^{\ell_j} \partial_x \theta_{j,n}e^{-\beta_n^{1/2}(\ell_j - x)}dx. \tag{3.59}$$

The right hand side of (3.59) converges to zero since $\partial_x \theta_{j,n}$ converges to zero and $e^{-\beta_n^{1/2}(\ell_j - x)}$ is bounded. Performing integration by parts to the left hand side, we obtain

$$\int_0^{\ell_j} \beta_n^{3/2}e^{-\beta_n^{1/2}(\ell_j - x)}u_{j,n}(x)dx - \frac{1}{\beta_n^{1/2}}\int_0^{\ell_j} e^{-\beta_n^{1/2}(\ell_j - x)}\partial_x^4 u_{j,n}(x)dx$$

$$= -\frac{1}{\beta_n^{1/2}}\partial_x^3 u_{j,n}(x)e^{-\beta_n^{1/2}(\ell_j - x)}\Big|_{x=0}^{x=\ell_j} + \partial_x^2 u_{j,n}(x)e^{-\beta_n^{1/2}(\ell_j - x)}\Big|_{x=0}^{x=\ell_j}$$

$$- \beta_n^{1/2}\partial_x u_{j,n}(x)e^{-\beta_n^{1/2}(\ell_j - x)}\Big|_{x=0}^{x=\ell_j} + \beta_n \, u_{j,n}(x)e^{-\beta_n^{1/2}(\ell_j - x)}\Big|_{x=0}^{x=\ell_j}.$$

It follows that

$$-\frac{1}{\beta_n^{1/2}}\partial_x^3 u_{j,n}(\ell_j) + \partial_x^2 u_{j,n}(\ell_j) - \beta_n^{1/2}\partial_x u_{j,n}(\ell_j) + \beta_n u_{j,n}(\ell_j) \longrightarrow 0.$$

We have used the fact that $\forall \alpha \in \mathbb{R}, \forall k \in \{0, 1, 2, 3\}$, $\beta_n^\alpha \partial_x^k u_{j,n}(0)e^{-\beta_n^{1/2}\ell_j}$ tend to 0.

With the same manner, by considering $\frac{1}{\beta_n^{1/2}}e^{-\beta_n^{1/2}x}$ instead of $\frac{1}{\beta_n^{1/2}}e^{-\beta_n^{1/2}(\ell_j - x)}$, we obtain that

$$\frac{1}{\beta_n^{1/2}}\partial_x^3 u_{j,n}(0) + \partial_x^2 u_{j,n}(0) + \beta_n^{1/2}\partial_x u_{j,n}(0) + \beta_n u_{j,n}(0) \longrightarrow 0.$$

In conclusion, for every inner node a_k,

$$\frac{1}{\beta_n^{1/2}}d_{kj}\partial_x^3 u_{j,n}(a_k) + \varepsilon\partial_x^2 u_{j,n}(a_k) + \beta_n^{1/2}d_{kj}\partial_x u_{j,n}(a_k) + \varepsilon\beta_n u_{j,n}(a_k) \longrightarrow 0.$$
(3.60)

with $\varepsilon \in \{-1, 1\}$. Summing over $j \in J_k$, this leads to

$$\partial_x^2 u_{j,n}(a_k) + \beta_n u_{j,n}(a_k) \longrightarrow 0.$$

We have used the conditions (3.18) and (3.32), the fact that $\left\|\frac{\partial_x \theta_{j,n}}{\beta_n^{1/2}}\right\|_\infty$ converges to zero, and the continuity of $u_{j,n}$ and $\partial_x^2 u_{j,n}$ at a_k. Going back to (3.60), we get

$$\frac{1}{\beta_n^{1/2}}d_{kj}\partial_x^3 u_{j,n}(a_k) + \beta_n^{1/2}d_{kj}\partial_x u_{j,n}(a_k) = \lambda_n^j \longrightarrow 0.$$
(3.61)

We fix j in J^{te} and a_k as an interior end of e_j. Then, we have the following three inequalities:

$$Re\left(\partial_x^3 u_{j,n}(a_k)\partial_x\overline{u_{j,n}}(a_k)\right) = Re\left(\frac{\partial_x^3 u_{j,n}(a_k)}{\beta_n^{1/2}}\beta_n^{1/2}\partial_x\overline{u_{j,n}}(a_k)\right)$$
$$= Re\left((\lambda_n^j - \beta_n^{1/2}\partial_x u_{j,n}(a_k))\beta_n^{1/2}\partial_x\overline{u_{j,n}}(a_k)\right)$$
$$\le -\beta_n\left|\partial_x\overline{u_{j,n}}(a_k)\right|^2 + \frac{a}{2}\beta_n\left|\partial_x\overline{u_{j,n}}(a_k)\right|^2 + \frac{1}{2a}\left|\lambda_n^j\right|^2,$$

$$Re\left(\partial_x\theta_{j,n}(a_k)\partial_x\overline{u_{j,n}}(a_k)\right) \le \frac{b}{2}\beta_n\left|\partial_x u_{j,n}(a_k)\right|^2 + \frac{1}{2b}\frac{\left|\partial_x\theta_{j,n}(a_k)\right|^2}{\beta_n}$$

and

$$-Re\left(i\beta_n f_{j,n}(a_k)u_{j,n}(a_k)\right) \leq \frac{c}{2}\beta_n^2 \left|u_{j,n}(a_k)\right|^2 + \frac{1}{2c}\left|f_{j,n}(a_k)\right|^2$$

for any positive numbers a, b, and c.

Going back to (3.58),

$$-\frac{1}{2}\beta_n^2 \left|u_{j,n}(a_k)\right|^2 + Re\left(\partial_x^3 u_{j,n}(a_k)\partial_x\overline{u_{j,n}}(a_k)\right) - \frac{1}{2}\left|\partial_x^2 u_{j,n}(a_k)\right|^2$$

$$- Re\left(\partial_x\theta_{j,n}(a_k)\partial_x\overline{u_{j,n}}(a_k)\right) - Re\left(i\beta_n f_{j,n}(a_k)\overline{u_{j,n}}(a_k)\right)$$

$$-\frac{1}{2a}\left|\lambda_n^j\right|^2 - \frac{1}{2c}\left|f_{j,n}(a_k)\right|^2 - \frac{1}{2b}\frac{\left|\partial_x\theta_{j,n}(a_k)\right|^2}{\beta_n}$$

$$\leq \left(-\frac{1}{2}+\frac{c}{2}\right)\beta_n^2\left|u_{j,n}(a_k)\right|^2 + \left(-1+\frac{a}{2}+\frac{b}{2}\right)\beta_n\left|\partial_x u_{j,n}(a_k)\right|^2 - \frac{1}{2}\left|\partial_x^2 u_{j,n}(a_k)\right|^2$$

$$\leq 0$$

for $a = b = c = \dfrac{1}{4}$. Using that

$$-\frac{1}{2a}\left|\lambda_n^j\right|^2 - \frac{1}{2c}\left|f_{j,n}(a_k)\right|^2 - \frac{1}{2d}\frac{\left|\partial_x\theta_{j,n}(a_k)\right|^2}{\beta_n} \longrightarrow 0$$

we deduce

$$\beta_n^2\left|u_{j,n}(a_k)\right|^2 \longrightarrow 0, \quad \left|\partial_x^2 u_{j,n}(a_k)\right|^2 \text{ and } \beta_n\left|\partial_x u_{j,n}(a_k)\right|^2 \longrightarrow 0.$$

Moreover, all the expressions

$$Re\left(\partial_x^3 u_{j,n}(a_k)\partial_x\overline{u_{j,n}}(a_k)\right), \ Re\left(\partial_x\theta_{j,n}(a_k)\partial_x\overline{u_{j,n}}(a_k)\right) \text{ and } Re\left(i\beta_n f_{j,n}(a_k)\overline{u_{j,n}}(a_k)\right)$$

converge to 0.

Now, let an elastic edge e_j be attached to a thermoelastic one at an internal node a_k. Take $q = x$ or $q = \ell_j - x$ in (3.56). By using the continuity conditions of $u_{j,n}$ and $\partial_x^2 u_{j,n}$ at a_k and the damping conditions (3.18) and (3.32), we have

$$\int_0^{\ell_j}\left(\left|\partial_x^2 u_{j,n}(x)\right|^2 + 3\beta_n^2\left|u_{j,n}(x)\right|^2\right)dx - 2Re\left(\partial_x^2 u_{j,n}(x)\partial_x\overline{u_{j,n}}(x)\Big|_{x=0}^{x=\ell_j}\right) \longrightarrow 0.$$

$$(3.62)$$

Similarly, taking $q = 1$ in (3.56) we obtain

$$-\frac{1}{2}\beta_n^2 \left|u_{j,n}(a_s)\right|^2 - \frac{1}{2}\left|\partial_x^2 u_{j,n}(a_s)\right|^2$$

$$+ Re\left(\partial_x^3 u_{j,n}(a_s)\partial_x\overline{u_{j,n}}(a_s)\right) - Re\left(\mathbf{i}\beta_n f_{j,n}(a_s)\overline{u}_{j,n}(a_s)\right) \longrightarrow 0, \qquad (3.63)$$

where a_s is the second end of e_j.

As for a thermoelastic edge we prove that all the expressions in the left hand side of (3.63) converge to zero, then

$$\int_0^{\ell_j} \left(\left|\partial_x^2 u_{j,n}(x)\right|^2 + 3\beta_n^2 \left|u_{j,n}(x)\right|^2\right) dx \longrightarrow 0. \qquad (3.64)$$

We iterate such procedure in each maximal subgraph of elastic edges of $\mathcal{G}$ to obtain (3.64) for all j in J^e.

In summary, we have $\|y_n\|_{\mathcal{H}} \longrightarrow 0$. This result contradicts the hypothesis that y_n has the unit norm.   □

We can now state the main result of this chapter.

Theorem 3.6 *The C_0-semigroup $S(t)$, generated by the operator $\mathcal{A}$, is exponentially stable. More precisely, the energy of the whole system (3.1)–(3.14) decays exponentially to zero.*

Proof The proof is a direct consequence of Lemmas 3.4 and 3.5.   □

3   Comment

If we replace the continuity condition of $\underline{\theta}$ at inner nodes,

$$\theta_j(a_k) = \theta_l(a_k) \quad j, l \in J^{te}(a_k), \quad a_k \in \mathcal{V}_{int}$$

and the condition

$$\sum_{j \in J^{te}(a_k)} d_{kj}(\gamma_j u_{xt}(a_k, t) - \theta_{j,x}(a_k, t)) = 0, \quad a_k \in \mathcal{V}_{int}$$

by the following:

$$\theta_j(a_k) = 0 \quad j \in J^{te}(a_k), \quad a_k \in \mathcal{V}_{int}$$

and Kirchhoff's law,

$$\sum_{j \in J^{te}(a_k)} d_{kj}\theta_{j,x}(a_k, t) = 0, \quad a_k \in \mathcal{V}_{int}$$

then we obtain the same results.

Furthermore, if we consider the following boundary conditions:

$$u_j(a_k, t) = 0, \ u_{j,x}(a_k, t) = 0, \qquad j \in J_k, \ a_k \in \mathcal{V}_{ext},$$
$$\theta_j(a_k, t) = 0, \qquad j \in J_k, \ a_k \in \mathcal{V}'_{ext},$$

$$u_j(a_k, t) = u_l(a_k, t) \qquad j, l \in J_k, \ a_k \in \mathcal{V}_{int},$$
$$\theta_j(a_k, t) = \theta_l(a_k, t) \qquad j, l \in J_k^{te}, \ a_k \in \mathcal{V}_{int},$$
$$u_{j,x}(a_k, t) = u_{l,x}(a_k, t) \qquad j, l \in J_k, \ a_k \in \mathcal{V}_{int},$$

$$\sum_{j \in J_k^{te}} d_{kj}\left(u_{j,xx}(a_k, t) - \gamma_j\theta_j(a_k, t)\right) = 0, \quad a_k \in \mathcal{V}_{int},$$

$$\sum_{j \in J_k^{te}} d_{kj}\left(u_{j,xxx}(a_k, t) - \gamma_j\theta_{j,x}(a_k, t)\right) + \sum_{j \in J_k^e} d_{kj}u_{j,xxx}(a_k, t) = 0, \quad a_k \in \mathcal{V}_{int},$$

$$\sum_{j \in J_k^{te}} d_{kj}\theta_{j,x}(a_k, t) = 0, \quad a_k \in \mathcal{V}_{int},$$

we can prove that the energy of the system decays exponentially to zero.

Chapter 4
Stability of a Tree-Shaped Network of Strings and Beams

In past decades, the dynamic behavior of networks of flexible structures has been studied by some authors, see, for instance, [7, 18, 49] and the references therein. The importance of these studies lies in the need for engineering to eliminate vibrations in such composite structures. In this chapter, we consider a model (S) of a tree-shaped network of N elastic materials, constituting *strings* and Euler Bernoulli *beams*.

Suppose that the equilibrium position of the tree of elastic strings and beams coincides with the tree $\mathcal{T}$ of N edges, $e_1, \ldots, e_N$ and $p = N + 1$ vertices, $a_1, \ldots, a_p$; and recall that $I = \{1, \ldots, p\}$ and $J = \{1, \ldots, N\}$ denote, respectively, the set of indices of the vertices and the set of indices of edges. We suppose that a_1 is the root of $\mathcal{T}$, and that a_1 and a_2 are ends of e_1.

Our model is then described as follows: Every string e_j satisfies the following equation:

$$u_{j,tt} - u_{j,xx} = 0 \text{ in } (0, \ell_j) \times (0, \infty), \tag{4.1}$$

and every beam e_j satisfies the following equation:

$$u_{j,tt} + u_{j,xxxx} = 0 \text{ in } (0, \ell_j) \times (0, \infty), \tag{4.2}$$

where $u_j = u_j(x, t)$ is the function describing the displacement of the string or beam e_j.
The initial conditions are

$$u_j(x, 0) = u_j^0(x), \quad u_{j,t}(x, 0) = u_j^1(x). \tag{4.3}$$

Denote by J_k^s (resp. J_k^b), the set of indices strings (resp. beams) adjacent to a_k and by $\mathcal{V}_{ext}^s$ and $\mathcal{V}_{ext}^b$, respectively, the set of external nodes of strings and those of beams, different from a_1. Then, the transmission conditions at the inner nodes are

$$
\begin{cases}
u_j(a_k, t) = u_l(a_k, t), & j, l \in J_k, \quad a_k \in \mathcal{V}_{int}, \\
u_{j,xx}(a_k, t) = u_{l,xx}(a_k, t), & j, l \in J_k^b, \quad a_k \in \mathcal{V}_{int}, \\
\displaystyle\sum_{j \in J_k^b} d_{kj} u_{j,x}(a_k, t) = 0, & a_k \in \mathcal{V}_{int}, \\
\displaystyle\sum_{j \in J_k^b} d_{kj} u_{j,xxx}(a_k, t) - \sum_{j \in J_k^s} d_{kj} u_{j,x}(a_k, t) = 0, & a_k \in \mathcal{V}_{int},
\end{cases} \tag{4.4}
$$

and the boundary conditions are

$$
\begin{cases}
u_{j_k}(a_k, t) = 0, & a_k \in \mathcal{V}_{ext}, \\
u_{j_k,xx}(a_k, t) = 0, & a_k \in \mathcal{V}_{ext}^b,
\end{cases} \tag{4.5}
$$

where j_k is the index of the unique edge adjacent to $a_k \in \mathcal{V}_{ext}$.

For a classical solution u of $(\mathcal{S})$, the energy is defined as the sum of the energy of its components, that is,

$$
E(t) = \frac{1}{2} \sum_{j=1}^{N} \int_0^{\ell_j} |u_{j,t}(x, t)|^2 \, dx + \frac{1}{2} \sum_{j \in J^s} \int_0^{\ell_j} |u_{j,x}(x, t)|^2 \, dx
$$

$$
+ \frac{1}{2} \sum_{j \in J^b} \int_0^{\ell_j} |u_{j,xx}(x, t)|^2 \, dx,
$$

where J^s and J^b are the respective index sets of strings and beams in the tree. Differentiate formally the energy function with respect to time t, we get

$$
\frac{dE}{dt}(t) = 0,
$$

and the system is conservative.

Stability of such models of networks of strings or of beams has been proved before, by applying a control at an external node or by forcing the damping conditions at inner nodes. In [4], the authors prove the polynomial stability of a star-shaped network of strings when a feedback is applied at the common node, and in [10] and [94], the authors prove a similar result for a tree of strings when the feedback is applied at an external node. In [3], we considered a network of beams. See also [100] for exponential stability of a star-shaped network of beams and [35] for asymptotic stability of a star-shaped network of Timoshenko beams. In [88] and [89] we added thermoelastic edges to the network of elastic materials to obtain an exponential stability result.

For strings-beams networks, see [5] where the authors considered a star-shaped network of beams and a string, with controls applied at all the exterior nodes. They proved a result of exponential stability. Some results of polynomial stability have

Fig. 4.1 First tree

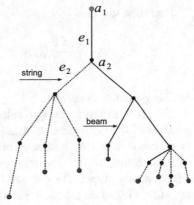

Fig. 4.2 Second tree

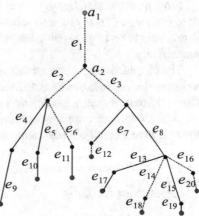

proved before, for coupled string-beam systems [11] (see [7] for general setting) and for chains of alternated beams and strings [12], when feedbacks are applied at inner nodes. The case of a $2 - d$ coupled system of a wave equation and a plate equation has been studied by Ammari and Nicaise in [6]. They proved a result of exponential stability under some geometric conditions.

In this chapter, we study a more general case of networks, in fact, it is the model $(\mathcal{S})$ presented at the beginning, stabilized by applying feedbacks at all leaves (the root remains free): Figs. 4.1 and 4.2. For this, let $\mathcal{V}^* = \mathcal{V} - \{a_1\}$, $\mathcal{V}_{ext}^* = \mathcal{V}_{ext} - \{a_1\}$ and let δ in $\{0, 1\}$ with $\delta = 1$ if e_1 is a string and $\delta = 0$ if e_1 is a beam. Then instead of (4.5) we take

$$u_1(a_1, t) = 0, \quad (1 - \delta)u_{1,xx}(a_1, t) = 0,$$

$$u_{j_k,x}(a_k, t) = -d_{kj}u_{j_k,t}(a_k, t), \quad a_k \in \mathcal{V}_{ext}^s,$$

$$u_{j_k,xxx}(a_k, t) = d_{kj}u_{j_k,t}(a_k, t), \quad u_{j_k,x}(a_k, t) = 0, \quad a_k \in \mathcal{V}_{ext}^b,$$

where some derivative boundary feedbacks are applied at external nodes (except the root).

Formally, we have

$$\frac{dE}{dt}(t) = - \sum_{a_k \in \mathcal{V}_{ext}^*} \left| u_{j_k,t}(a_k, t) \right|^2 \leq 0.$$

So the system is dissipative.

We prove different decay results of the energy of the system depending on the position of beams relative to strings. Precisely, $(\mathcal{S})$ is exponentially stable if there is no beam following a string from the root to leaves and polynomially stable if not. Moreover, we give an example corresponding to the last case which is not exponentially stable. The method that we use to show exponential or polynomial stability is based on the resolvent approach.

From now, for simplicity, we use sometimes the following two **notations**: $u_n = o(v_n)$ and $u_n = O(v_n)$ for two sequences of complex numbers u_n and v_n where v_n is non-zero for n large enough; they mean $\frac{u_n}{v_n} \longrightarrow 0$ as $n \longrightarrow \infty$ and $\frac{u_n}{v_n}$ is bounded, respectively

The chapter is organized as follows: In Sect. 1, we reformulate the system $(\mathcal{S})$ as an evolution equation in a Hilbert space and prove that it is associated with a C_0-semigroup of contractions, and in Sect. 2, by using frequency domain method, we first prove, under some conditions, that the system $(\mathcal{S})$ is exponentially stable, then we give a result of polynomial stability.

1   Abstract Setting

First, Recall that J^s and J^b are the respective sets of indices of strings and beams in the tree. Then, for a function f on $\mathcal{T}$, we set $f^s = (f_j)_{j \in J^s}$ and $f^b = (f_j)_{j \in J^b}$, and we rewrite f as $f = (f^s, f^b)$.

The aim of this section is to rewrite the system $(\mathcal{S})$ as an evolution equation in an appropriate Hilbert space. We then prove the existence and uniqueness of solutions of the problem using semigroup theory.

Let us consider

$$V = \left\{ \underline{f} = (f^s, f^b) \in \prod_{j \in J^s} H^1(0, \ell_j) \times \prod_{j \in J^b} H^2(0, \ell_j) \mid \underline{f} \text{ satisfies (4.6)} \right\},$$

where

$$\begin{cases} f_1(a_1) = 0, \\ f_j(a_k) = f_l(a_k), \quad j, l \in J_k, \ a_k \in \mathcal{V}_{int}, \\ \partial_x f_{j_k}(a_k) = 0, \quad a_k \in \mathcal{V}_{ext}^b, \\ \sum_{j \in J^b(a_k)} d_{kj} \partial_x f_j(a_k) = 0, \quad a_k \in \mathcal{V}_{int}. \end{cases} \tag{4.6}$$

Note that we can rewrite the last two equations in one, as follows:

$$\sum_{j \in J_k^b} d_{kj} \partial_x f_j(a_k) = 0, \quad a_k \in V^*.$$

Define the energy space of $(\mathcal{S})$ by

$$\mathcal{H} = V \times \prod_{j=1}^N L^2(0, \ell_j)$$

endowed by the inner product

$$\langle y, \tilde{y} \rangle_{\mathcal{H}} := \sum_{j \in J^s} \left\langle \partial_x f_j, \partial_x \tilde{f}_j \right\rangle + \sum_{j \in J^b} \left\langle \partial_x^2 f_j, \partial_x^2 \tilde{f}_j \right\rangle + \sum_{j=1}^N \langle g_j, \tilde{g}_j \rangle,$$

where $y = (f, g)$, and $\tilde{y} = (\tilde{f}, \tilde{g})$, Then, $\mathcal{H}$ is a Hilbert space.

Now define the operator $\mathcal{A}$ in $\mathcal{H}$ by

$$\mathcal{D}(\mathcal{A}) = \left\{ \begin{array}{c} y = (u_s, u_b, v_s, v_b) \in V \times V \mid u_s \in \prod_{j \in J^s} H^2(0, \ell_j), u_b \in \prod_{j \in J^b} H^4(0, \ell_j) \\ \text{and } y \text{ satisfies (4.7)} \end{array} \right\},$$

where

$$\begin{cases} \partial_x u_{j_k}(a_k) = -d_{kj} v_{j_k}(a_k), & a_k \in V_{ext}^s, \\ (1 - \delta) \partial_x^2 u_1(a_1) = 0, \\ \partial_x^2 u_j(a_k) = \partial_x^2 u_l(a_k), & j, l \in J_k^b, \ a_k \in V_{int}, \\ \partial_x^3 u_{j_k}(a_k) = d_{kj} v_{j_k}(a_k), & a_k \in V_{ext}^b, \\ \sum_{j \in J_k^s} d_{kj} \partial_x u_j(a_k) - \sum_{j \in J_k^b} d_{kj} \partial_x^3 u_j(a_k) = 0, & a_k \in V_{int}, \end{cases} \tag{4.7}$$

and

$$\mathcal{A} \begin{pmatrix} u_s \\ u_b \\ v_s \\ v_b \end{pmatrix} = \begin{pmatrix} v_s \\ v_b \\ \partial_x^2 u_s \\ -\partial_x^4 u_b \end{pmatrix}, \quad (u_s, u_b, v_s, v_b) \in \mathcal{D}(\mathcal{A}).$$

Then, putting $y = (\underline{u}, \underline{u}_t)$, we write the system $(\mathcal{S})$ into the following first-order evolution equation:

$$\begin{cases} \frac{dy}{dt} = \mathcal{A}y, \\ y(0) = y^0, \end{cases} \tag{4.8}$$

on the energy space $\mathcal{H}$, where $y_0 = (u^0, u^1)$.

We have the following result:

Lemma 4.1 *The operator $\mathcal{A}$ is the infinitesimal generator of a C_0-semigroup of contractions $(T(t))_{t \geq 0}$.*

Proof By Lumer–Phillips Theorem (Theorem 1.15), it suffices to show that $\mathcal{A}$ is m-dissipative. First, for any $y \in \mathcal{D}(\mathcal{A})$ we have

$$Re(\langle \mathcal{A}y, y \rangle_{\mathcal{H}}) = Re \left(\sum_{j \in J^s} \int_0^{\ell_j} (\partial_x v_j \partial_x \overline{u}_j dx + \partial_x^2 u_j \overline{v}_j) dx \right.$$

$$\left. + \sum_{j \in J^b} \int_0^{\ell_j} (\partial_x^2 v_j \partial_x^2 \overline{u}_j dx - \partial_x^4 u_j \overline{v}_j) dx \right).$$

Using integration by parts, we obtain using, boundary and transmission conditions (4.6) and (4.7),

$$Re(\langle \mathcal{A}y, y \rangle_{\mathcal{H}}) = - \sum_{a_k \in \mathcal{V}_{ext}^*} \left| v_{j_k}(a_k) \right|^2 \leq 0.$$

Then the operator $\mathcal{A}$ is dissipative.

We show now that every positive real number λ belongs to $\rho(\mathcal{A})$. Let $z = (\underline{f}, \underline{g}) \in \mathcal{H}$, we look for $y = (\underline{u}, \underline{v}) \in \mathcal{D}(\mathcal{A})$ such that

$$(\lambda - \mathcal{A})y = z,$$

i.e.,

$$\lambda u_j - v_j = f_j, \ j = 1, \ldots, N, \tag{4.9}$$

$$\lambda v_j - \partial_x^2 u_j = g_j, \ j \in J^s, \tag{4.10}$$

$$\lambda v_j + \partial_x^4 u_j = g_j, \ j \in J^b. \tag{4.11}$$

Then,

$$\lambda^2 u_j - \partial_x^2 u_j = g_j + \lambda f_j, \quad j \in J^s, \tag{4.12}$$

$$\lambda^2 u_j + \partial_x^4 u_j = g_j + \lambda f_j, \quad j \in J^b. \tag{4.13}$$

Let $\underline{w}$ in V. Multiplying the first equation by w_s and the second equation by w_b, and integrating by parts, we get, respectively,

$$\lambda^2 \int_0^{\ell_j} u_j \overline{w_j} dx + \int_0^{\ell_j} \partial_x u_j \partial_x \overline{w_j} dx - \partial_x u_j \overline{w_j} \Big|_0^{\ell_j} = \int_0^{\ell_j} (g_j + \lambda f_j) \overline{w_j} dx,$$

for j in J^s and

$$\lambda^2 \int_0^{\ell_j} u_j \overline{w_j} dx + \int_0^{\ell_j} \partial_x^2 u_j \partial_x^2 \overline{w_j} dx + \partial_x^3 u_j \overline{w_j} \Big|_0^{\ell_j} - \partial_x^2 u_j \partial_x \overline{w_j} \Big|_0^{\ell_j}$$

$$= \int_0^{\ell_j} (g_j + \lambda f_j) \overline{w_j} dx,$$

for j in J^b. Then, summing the two obtained equations, the left hand side will be

$$\lambda^2 \sum_{j=1}^N \int_0^{\ell_j} u_j \overline{w_j} dx + \sum_{j \in J^s} \int_0^{\ell_j} \partial_x u_j \partial_x \overline{w_j} dx + \sum_{j \in J^b} \int_0^{\ell_j} \partial_x^2 u_j \partial_x^2 \overline{w_j} dx$$

$$+ \sum_{a_k \in V_{int}} \left(- \sum_{j \in J_k^s} d_{kj} \overline{w_j}(a_k) \partial_x u_j(a_k) + \sum_{j \in J_k^b} d_{kj} \overline{w_j}(a_k) \partial_x^3 u_j(a_k) \right)$$

$$+ \sum_{a_k \in V_{ext}} v_{j_k}(a_k) \overline{w_{j_k}}(a_k) - \sum_{a_k \in V_{int}} \sum_{j \in J_k^b} d_{kj} \partial_x^2 u_j(a_k) \partial_x \overline{w_j}(a_k).$$

We find, by taking into account (4.6) and (4.7)

$$a(\underline{u}, \underline{w}) = F(\underline{w}), \tag{4.14}$$

where

$$a(\underline{u}, \underline{w}) = \lambda^2 \sum_{j=1}^N \int_0^{\ell_j} u_j \overline{w_j} dx + \sum_{j \in J^s} \int_0^{\ell_j} \partial_x u_j \partial_x \overline{w_j} dx + \sum_{j \in J^b} \int_0^{\ell_j} \partial_x^2 u_j \partial_x^2 \overline{w_j} dx$$

$$+ \lambda \sum_{a_k \in V_{ext}} u_{j_k}(a_k) \overline{w_{j_k}}(a_k)$$

and

$$F(\underline{w}) = \sum_{j=1}^{N} \int_0^{\ell_j} (g_j + \lambda f_j)\overline{w_j}\,dx + \lambda \sum_{a_k \in \mathcal{V}_{ext}} f_{jk}(a_k)\overline{w_{jk}}(a_k).$$

a is a continuous sesquilinear form on $V \times V$, and F is a continuous anti-linear form on V. Moreover, there exists $C > 0$ such that, for every $\underline{w} \in V$,

$$\left| a(\underline{w}, \underline{w}) \right| \geq C \left\| \underline{w} \right\|_V^2,$$

where

$$\left\| \underline{u} \right\|_V^2 = \sum_{j=1}^{N} \int_0^{\ell_j} \left| u_j \right|^2 dx + \sum_{j \in J^s} \int_0^{\ell_j} \left| \partial_x u_j \right|^2 dx + \sum_{j \in J^b} \int_0^{\ell_j} \left| \partial_x^2 u_j \right|^2 dx.$$

By the Lax–Milgram's lemma, problem (4.14) has a unique solution $\underline{u}$ in V. It is easy to verify that: $\underline{u}$ belongs to $\prod_{j \in J^s} H^2(0, \ell_j) \times \prod_{j \in J^b} H^4(0, \ell_j)$, $\underline{v} = \lambda \underline{u} - \underline{f} \in V$, u_s and u_b satisfies, respectively, (4.10) and (4.11), and the conditions

$$\begin{cases} \partial_x u_{jk}(a_k) = -d_{kj} v_{jk}(a_k), & a_k \in \mathcal{V}_{ext}^s, \\ (1 - \delta)\partial_x^2 u_1(\ell_1) = 0, \\ \partial_x^2 u_j(a_k) = \partial_x^2 u_l(a_k), & j, l = J_k^b, \ a_k \in \mathcal{V}_{int}, \\ \partial_x^3 u_{jk}(a_k) = d_{kj} v_{jk}(a_k), & a_k \in \mathcal{V}_{ext}^b, \\ \sum_{j \in J_k^s} d_{kj} \partial_x u_j(a_k) - \sum_{j \in J_k^b} d_{kj} \partial_x^3 u_j(a_k) = 0, & a_k \in \mathcal{V}_{int}. \end{cases}$$

Furthermore,

$$\left\| y \right\|_{\mathcal{H}}^2 \leq c \left\| z \right\|_{\mathcal{H}}^2,$$

where c is a positive constant independent of y. In conclusion, $y = (\underline{u}, \underline{v}) \in \mathcal{D}(\mathcal{A})$ and $(\lambda - \mathcal{A})^{-1} \in \mathcal{L}(\mathcal{H})$, that is, $\lambda \in \rho(\mathcal{A})$. □

Corollary 4.2 *For an initial datum* $y^0 \in \mathcal{H}$, *there exists a unique solution* $y \in C([0, +\infty), \mathcal{H})$ *of problem* (4.8). *Moreover, if* $y^0 \in \mathcal{D}(\mathcal{A})$, *then* $y \in C([0, +\infty), \mathcal{D}(\mathcal{A})) \cap C^1([0, +\infty), \mathcal{H})$.

Remark 4.3 Note that, by the Sobolev embedding theorem, we deduce that the canonical injection $\mathcal{D}(\mathcal{A}) \hookrightarrow \mathcal{H}$ is compact, then the operator $\mathcal{A}$ has compact resolvent (Proposition 1.5). Hence, the spectrum of $\mathcal{A}$ consists of all isolated eigenvalues, i.e., $\sigma(\mathcal{A}) = \sigma_p(\mathcal{A})$.

2 Asymptotic Behavior

The aim of this section is to show that the system (S) is asymptotically stable. Moreover, we will prove that the solution is exponentially stable if there is no beam following a string, from the root to the leaves, as in the first tree (Fig. 4.1), and polynomially stable if at least a beam follows a string, as in the second tree (Fig. 4.2). Finally, the lack of exponential stability is proved on an example.

For the sequel, we need some definitions and notations. We say that a string verifies the NFB property if it is not followed by any beam, and such string is called a NFB-string. A beam satisfies the NFS property if it is not followed by any string and is not attached simultaneously to a string and a beam that is followed by a string; it is called a NFS-beam.

The NFB-strings form the first layer of strings; it may be empty, and then we remove it. The first layer of beams is formed of NFS-beams in the new tree $\mathcal{T}_1$. The second layer of strings is composed of NFB-strings of the tree $\mathcal{T}_2$ obtained from $\mathcal{T}_1$ by removing the NFS-beams, etc. Let r be the number of layers of beams in $\mathcal{T}$ not containing e_1. Note that $r \geq 1$.

Returning back to the tree in Fig. 4.2; the NFB-strings of $\mathcal{T}$ are grouped into two connected components $\{e_{12}\}$, and $\{e_{14}, e_{18}, e_{16}, e_{20}\}$. Then the first layer of beams is the union of its three connected components $\{e_4, e_9, e_5, e_{10}\}$, $\{e_{11}\}$, and $\{e_{13}, e_{17}, e_{15}, e_{19}\}$.

For a subgraph $\mathcal{G}$ of $\mathcal{T}$, we denote by $\mathcal{V}_{ext}(\mathcal{G})$ the set of the endpoints of $\mathcal{G}$, except the outer node (the nearest node of $\mathcal{G}$ to the root), and by $\mathcal{V}_{int}(\mathcal{G})$ the set of the inner nodes, except the outer node. Finally, if a_k is a node of $\mathcal{G}$, then we denote by $J_k(\mathcal{G})$ the index set of edges belonging to $\mathcal{G}$ and attached to a_k.

2.1 Asymptotic Stability

In this section, we prove condition (1.9) in Theorem 1.25, then the system (S) is asymptotically stable.

Theorem 4.4 *The semigroup $(T(t))_{t \geq 0}$, generated by the operator $\mathcal{A}$ is asymptotically stable.*

Proof Using Corollary 1.22 it suffices to show that $i\mathbb{R} \subset \rho(\mathcal{A})$; otherwise, by taking into account Remark 4.3, there is a real number β, such that $\lambda := i\beta$ is an eigenvalue of $\mathcal{A}$. Let $y = (\underline{u}, \underline{v})$ the corresponding eigenvector. We have

$$\begin{cases} v_j = \lambda u_j & \text{for } j \text{ in } \{1, \dots, N\}, \\ \partial_x^2 u_j = \lambda v_j & \text{for } j \text{ in } J^s, \\ -\partial_x^4 u_j = \lambda v_j & \text{for } j \text{ in } J^b. \end{cases} \quad (4.15)$$

If $\lambda = 0$, multiplying the second and the third equations of (4.15) by u_j and summing, we obtain, using (4.6) and (4.7),

$$\sum_{j \in J^s} \left\| \partial_x u_j \right\|^2 + \sum_{j \in J^b} \left\| \partial_x^2 u_j \right\|^2 = 0$$

which implies that u_j is constant on e_j for every $j \in J^e$ and $\partial_x u_j$ is constant on e_j for every $j \in J^b$.

For $a_k \in \mathcal{V}_{ext}^b$, $\partial_x u_{j_k} = 0$, then u_{j_k} is constant on e_{j_k}. Moreover, using the fourth condition in (4.6) we deduce that u_j is constant on e_j for every $j \in J^b$. Hence, $\underline{u} = 0$, by the first and the second condition in (4.6). Using again (4.15), with $\lambda = 0$, we deduce that $\underline{v} = 0$. Thus $y = 0$.

In the sequel, we suppose that $\lambda \neq 0$. Taking the real part of the inner product of $\lambda y - \mathcal{A} y = 0$ with y in $\mathcal{H}$, we obtain

$$Re(\langle \mathcal{A} y, y \rangle_{\mathcal{H}}) = - \sum_{a_k \in \mathcal{V}_{ext}^*} \left| v_{j_k}(a_k) \right|^2 = 0.$$

Thus $v_{j_k}(a_k) = 0$ for $a_k \in \mathcal{V}_{ext}^*$ and then $u_{j_k}(a_k) = 0$ for $a_k \in \mathcal{V}_{ext}^*$, which holds for $k = 1$; $\partial_x^3 u_{j_k}(a_k) = 0$ for $a_k \in \mathcal{V}_{ext}^b$ and $\partial_x u_{j_k}(a_k) = 0$ for $a_k \in \mathcal{V}_{ext}^s$.

Then, u is zero on every edge attached to a leaf, and by iteration, on every maximal subgraph of strings not followed by beams.

Now let $\mathcal{G}$ be a maximal subgraph of beams not followed by strings. We want to prove that u is zero on $\mathcal{G}$.

First Case $\mathcal{G} = \mathcal{T}$. For each j in $\{1, \ldots, N\}$, substituting the first equation of (4.15) into the third, we obtain,

$$\partial_x^4 u_j + \lambda^2 u_j = 0. \tag{4.16}$$

Again, as in the previous chapter, we use a matrix method as follows (see page 40 for any explanation on the notations used):

To a function f on $\mathcal{T}$ is associated the matrix function F defined by

$$F : [0, 1] \longrightarrow \mathbb{C}^{p \times p}, x \longmapsto F(x) = (f_{jk}(x))_{p \times p},$$

with

$$f_{jk}(x) = e_{jk} f_{s(j,k)} \left[\ell_{s(j,k)} \left(\frac{1 + d_{js(j,k)}}{2} - x d_{js(j,k)} \right) \right].$$

The system (4.16) is then rewritten as

$$L^{(-4)} * U'''' + \lambda^2 U = 0. \tag{4.17}$$

Integrating equation (4.17), we obtain

$$U(x) = A_1 * \cos(\sqrt{\beta}Lx) + A_2 * \sin(\sqrt{\beta}Lx)$$
$$+ B_1 * \cosh(\sqrt{\beta}Lx) + B_2 * \sinh(\sqrt{\beta}Lx), \tag{4.18}$$

where without loss of generality we have supposed that $\beta > 0$ and with $A_1, A_2, B_1, B_2 \in \mathbb{C}^{p \times p}$. Then,

$$L^{(-1)} * U' = \sqrt{\beta}\left(-A_1 * \sin(\sqrt{\beta}Lx) + A_2 * \cos(\sqrt{\beta}Lx)\right.$$
$$\left. + B_1 * \sinh(\sqrt{\beta}x) + B_2 * \cosh(\sqrt{\beta}Lx)\right), \tag{4.19}$$

$$L^{(-2)} * U'' = \beta\left(-A_1 * \cos(\sqrt{\beta}Lx) - A_2 * \sin(\sqrt{\beta}Lx)\right.$$
$$\left. + B_1 * \cosh(\sqrt{\beta}x) + B_2 * \sinh(\sqrt{\beta}Lx)\right), \tag{4.20}$$

$$L^{(-3)} * U''' = \beta^{3/2}\left(A_1 * \sin(\sqrt{\beta}Lx) - A_2 * \cos(\sqrt{\beta}Lx)\right.$$
$$\left. + B_1 * \sinh(\sqrt{\beta}x) + B_2 * \cosh(\sqrt{\beta}Lx)\right). \tag{4.21}$$

The function U satisfies also,

$$U(1 - x) = U(x)^T. \tag{4.22}$$

The boundary and transmission conditions can be expressed as follows:

For the continuity condition of $\underline{u}$ at the inner nodes, there exists $\varphi = \begin{pmatrix} \varphi_1 \\ \vdots \\ \varphi_p \end{pmatrix} \in \mathbb{C}^p$

such that

$$U(0) = (\varphi e^T) * E, \tag{4.23}$$

where $e = \begin{pmatrix} 1 \\ \vdots \\ 1 \end{pmatrix} \in \mathbb{R}^p$.

Since $\underline{u}$ is zero at all external nodes, then $\varphi_k = 0$ when a_k is an external node.

The continuity condition of $\partial_x^2 \underline{u}$ at the interior nodes and the fact that $\partial_x^2 \underline{u}$ is zero at the root can be expressed in this manner, there exists $\psi = \begin{pmatrix} \psi_1 \\ \vdots \\ \psi_p \end{pmatrix} \in \mathbb{C}^p$ such that $\psi_1 = 0$, and

$$L^{(-2)} * U''(0) = (\psi e^T) * E. \tag{4.24}$$

The fourth condition of (4.6) and the fifth of (4.7) applied to $\underline{u}$ are expressed, respectively, as follows:

$$(L^{(-1)} * U'(0) * E^*)e = 0, \tag{4.25}$$

and

$$(L^{(-3)} * U'''(0) * E^*)e = 0, \tag{4.26}$$

where E^* is obtained from E by annulling the first line.

Substituting (4.23)–(4.24) in (4.19)–(4.21), and (4.25)-(4.26) in (4.18)–(4.20) leads to

$$A_1 = \frac{1}{2}(U(0) - L^{(-2)} * U''(0)) = \frac{1}{2}((\varphi - \psi)e^T) * E, \tag{4.27}$$

$$B_1 = \frac{1}{2}(U(0) + L^{(-2)} * U''(0)) = \frac{1}{2}((\varphi + \psi)e^T) * E, \tag{4.28}$$

and

$$(A_2 * E^*)e = 0, \tag{4.29}$$

$$(B_2 * E^*)e = 0. \tag{4.30}$$

By taking $x = 1$ in (4.18) and (4.20) and using (4.22), we get, by combining the two obtained equations,

$$\sinh^{(-1)}(\sqrt{\beta}L) * (B_1^T - B_1 \cosh(\sqrt{\beta}L)) = B_2. \tag{4.31}$$

Multiplying, in the Hadamard product, the above equation by E^*, we get, using (4.30)

$$\left(\sinh^{(-1)}(\sqrt{\beta}L) * (B_1^T - B_1 \cosh(\sqrt{\beta}L)) * E^*\right)e = 0. \tag{4.32}$$

We recall the following elementary rules for a matrix $M \in \mathbb{C}^{p \times p}$ (see [22]):

$$(M * B_1^T)e = M(\varphi + \psi), \quad (M * B_1)e = diag(Me)(\varphi + \psi). \tag{4.33}$$

Then, (4.32) implies

$$J(\varphi + \psi) = 0,$$

where

$$J = \sinh^{(-1)}(\sqrt{\beta}L) * E^* - diag\left[\left(\sinh^{(-1)}(\sqrt{\beta}L) * \cosh(\sqrt{\beta}L) * E^*\right)e\right].$$

The matrix, obtained from $J * E^{*T} * E^*$ by removing rows and columns that are zero, is a strictly diagonally dominant matrix. Since $\varphi_1 = \psi_1 = 0$, this implies that the vector $\varphi + \psi$, and hence, the matrix B_1, is zero. Returning to (4.31), we deduce that $B_2 = 0$.

For $j = 1, \ldots, N$, the expression of u_j is then

$$u_j(x) = a_1^j \cos(\sqrt{\beta}x) + a_2^j \sin(\sqrt{\beta}x), \quad \left(a_1^j, a_2^j \in \mathbb{C}\right),$$

which easily implies, using the transmissions conditions and the fact that $\underline{u}$ and $\partial_x \underline{u}$ vanish at the leaves, that

$$\underline{u} = 0.$$

Second Case $\mathcal{G} \neq \mathcal{T}$. Let a' be the nearest node of $\mathcal{G}$ to a_1. Then, a' is an end of at least one string.

For simplicity of notations, we will suppose, in this part, that $a' = a_1$ and $\mathcal{G} = \mathcal{T}$ but with boundary conditions at a_1:

$$\begin{cases} u_j(a_1) = u_l(a_1) \quad j, l \in J_1, \\ \partial_x^2 u_j(a_1) = \partial_x^2 u_l(a_1) \quad j, l \in J_1, \\ \sum_{j \in J_1^b} d_{1j} \partial_x u_j(a_1) = 0. \end{cases}$$

As for the first case, there is $\psi = \begin{pmatrix} \psi_1 \\ \vdots \\ \psi_p \end{pmatrix}$ in $\mathbb{C}^p$ with $\varphi_k = 0$ when $a_k \in \mathcal{V}_{ext}^*$,

such that

$$U(0) = (\varphi e^T) * E$$

and

$$L^{(-2)} * U''(0) = (\psi e^T) * E.$$

The third and fourth conditions of (4.6) and the fifth of (4.7) applied to $\underline{u}$ are expressed as follows:

$$(L^{(-1)} * U'(0) * E)e = 0$$

and

$$(L^{(-3)} * U'''(0) * E^*)e = 0.$$

As in the first case, we obtain (4.27) to (4.31). Moreover, $(\varphi_k + \psi_k)_{k=2,\ldots,p}$ will be the trivial solution of a homogeneous linear system whose matrix is invertible. Then, $B_1 * E^*$ and $B_2 * E^{*T} * E^*$ are zero.

For $j \in \{1, \ldots, N\} - J_1$, the expression of u_j is then

$$u_j(x) = a_1^j \cos(\sqrt{\beta}x) + a_2^j \sin(\sqrt{\beta}x), \quad \left(a_1^j, a_2^j \in \mathbb{C}\right),$$

which easily implies, by using the transmissions conditions and the fact that $\underline{u}$ and $\partial_x \underline{u}$ vanish at the leaves,

$$u_j = 0.$$

Then, we can suppose that $\mathcal{G}$ is a start of beams e_j, $j \in J_1$. Without loss of generality, we identify e_j with $(0, \ell_j)$ by taking $\pi_j(0) = a_1$. In such case, we have the following system:

$$\begin{cases} u_j(\ell_j) = 0, \ j \in J_1, \\ \partial_x u_j(\ell_j) = 0 \text{ and } \partial_x^3 u_j(\ell_j) = 0, \ j \in J_1, \\ u_j(0) = u^l(0) \text{ and } \partial_x^2 u_j(0) = \partial_x^2 u^l(0), \ j, l \in J_1, \\ \sum_j \partial_x u_j(0) = 0. \end{cases}$$

taking into account that $u_j(x)$ is of the form

$$u_j(x) = a_1^j \cos(\sqrt{\beta}x) + a_2^j \sin(\sqrt{\beta}x) + b_1^j \cosh(\sqrt{\beta}x) + b_2^j \sinh(\sqrt{\beta}x) \ \left(a_1^j, a_2^j, b_1^j, b_2^j \in \mathbb{C}\right),$$

This implies

$$\begin{cases} a_1^j \cos(\sqrt{\beta}\ell_j) + a_2^j \sin(\sqrt{\beta}\ell_j) + b_1^j \cosh(\sqrt{\beta}\ell_j) + b_2^j \sinh(\sqrt{\beta}\ell_j) = 0, \ j \in J_1, \\ -a_1^j \sin(\sqrt{\beta}\ell_j) + a_2^j \cos(\sqrt{\beta}\ell_j) = 0, \ j \in J_1, \\ b_1^j \sinh(\sqrt{\beta}\ell_j) + b_2^j \cosh(\sqrt{\beta}\ell_j) = 0, \ j \in J_1, \\ b_1^j = b_1^l, \ j, l \in J_1, \\ \sum_j a_2^j + b_2^j = 0. \end{cases}$$

The discriminant of the above system is

$$\Delta = \sum_j \left(\prod_{k \neq j} \cosh(\sqrt{\beta}\ell_k) \left(\sin(\sqrt{\beta}\ell_j) + \sinh(\sqrt{\beta}\ell_j) \right) \right),$$

which is different from zero. We conclude that $u_j = 0$, $j \in J_1$. That is, $\underline{u}$ is null on $\mathcal{G}$.

By iteration, and using transmission conditions, we conclude that $\underline{u}$ is zero on $\mathcal{T}$. The above discussion is sufficient to conclude that $y = 0$, which contradicts the fact that $y \neq 0$.   □

2.2 Exponential Stability

In this section, we suppose that there are no beam following a string (from the root to leaves), that is to say on every tree branch there is no beam between a string and a leaf (Fig. 4.1). We prove that the solution of the whole system $(\mathcal{S})$ is exponentially stable. Note that in this case, all strings are NFB and there is at most only and only maximal subgraph of beams (which necessarily contains e_1).

Theorem 4.5 *If there are no beam following a string, then the system $(\mathcal{S})$ is exponentially stable.*

Proof It suffices to prove that (1.10) holds. Suppose the conclusion is false. Then, there exists a sequence (β_n) of real numbers, without loss of generality, with $\beta_n \longrightarrow +\infty$, and a sequence of vectors $(y_n) = (\underline{u}_n, \underline{v}_n)$ in $\mathcal{D}(\mathcal{A})$ with $\|y_n\|_{\mathcal{H}} = 1$, such that

$$\|(i\beta_n I - \mathcal{A})y_n\|_{\mathcal{H}} \longrightarrow 0,$$

which is equivalent to

$$i\beta_n u_{j,n} - v_{j,n} = f_{j,n} \longrightarrow 0, \quad \text{in } H^1(0, \ell_j), \quad j \text{ in } J^s, \tag{4.34}$$

$$i\beta_n u_{j,n} - v_{j,n} = f_{j,n} \longrightarrow 0, \quad \text{in } H^2(0, \ell_j), \quad j \text{ in } J^b, \tag{4.35}$$

$$i\beta_n v_{j,n} - \partial_x^2 u_{j,n} = g_{j,n} \longrightarrow 0, \quad \text{in } L^2(0, \ell_j), \quad j \text{ in } J^s, \tag{4.36}$$

$$i\beta_n v_{j,n} + \partial_x^4 u_{j,n} = g_{j,n} \longrightarrow 0, \quad \text{in } L^2(0, \ell_j), \quad j \text{ in } J^b. \tag{4.37}$$

Then,

$$-\beta_n^2 u_{j,n} - \partial_x^2 u_{j,n} = g_{j,n} + i\beta_n f_{j,n}, \quad j \text{ in } J^s, \tag{4.38}$$

$$-\beta_n^2 u_{j,n} + \partial_x^4 u_{j,n} = g_{j,n} + i\beta_n f_{j,n}, \quad j \text{ in } J^b, \tag{4.39}$$

and

$$\|v_{j,n}\|^2 - \beta_n^2 \|u_{j,n}\|^2 \longrightarrow 0, \quad j = 1, \ldots, N. \tag{4.40}$$

First, since

$$Re(\langle(i\beta_n - \mathcal{A})y_n, y_n\rangle_{\mathcal{H}}) = \sum_{a_k \in \mathcal{V}_{ext}^*} \left|v_{j_k,n}(a_k)\right|^2,$$

we obtain

$$\left|v_{j_k,n}(a_k)\right| \longrightarrow 0, \text{ for } a_k \in \mathcal{V}_{ext}^*. \tag{4.41}$$

Next, we decompose the rest of the proof into two steps.

Step 1 We will prove that for every j in J^s

$$\|\partial_x u_{j,n}\|, \|v_{j,n}\| \longrightarrow 0. \tag{4.42}$$

We start by strings attached to external nodes. To do this let $a_k \in \mathcal{V}_{ext}^s$. Then, using (4.34), (4.41), and damping conditions at external nodes, we have

$$\beta_n u_{j_k,n}(a_k) \longrightarrow 0 \text{ and } \partial_x u_{j_k,n}(a_k) \longrightarrow 0. \tag{4.43}$$

Taking the real part of the inner product of (4.38) by $q_{j_k} \partial_x u_{j_k,n}$ with $q_{j_k} = x$ or $q_{j_k} = (\ell_{j_k} - x)$ yields

$$\frac{1}{2}\beta_n^2 \left|u_{j_k,n}(a_k)\right|^2 \Big|_0^{\ell_{j_k}} + \frac{1}{2} \left|\partial_x u_{j_k,n}(x)\right|^2 q_{j_k}(x) \Big|_0^{\ell_{j_k}}$$
$$- \frac{1}{2} \int_0^{\ell_{j_k}} \left(\left|\partial_x u_{j_k,n}(x)\right|^2 + \beta_n^2 \left|u_{j_k,n}(x)\right|^2\right) \partial_x q_{j_k}(x) dx$$
$$+ Re\left(i\beta_n f_{j_k,n}(x) q_{j_k}(x) \overline{u_{j_k,n}}(x)\right)\Big|_0^{\ell_{j_k}} \longrightarrow 0. \tag{4.44}$$

Using (4.43) in (4.44), $q_{j_k}(x) = x$ or $q_{j_k}(x) = \ell_{j_k} - x$, leads to

$$-\frac{1}{2} \int_0^{\ell_{j_k}} \left(\left|\partial_x u_{j_k,n}(x)\right|^2 + \beta_n^2 \left|u_{j_k,n}(x)\right|^2\right) dx \to 0.$$

Moreover, if we take $q_{j_k} = 1$, we obtain that

$$-\frac{1}{2}\beta_n^2 \left|u_{j_k,n}(a_{k'})\right|^2 - \frac{1}{2} \left|\partial_x u_{j_k,n}(a_{k'})\right|^2 - Re\left(i\beta_n f_{j_k,n}(a_{k'})\overline{u_{j_k,n}}(a_{k'})\right) \longrightarrow 0,$$

where $a_{k'}$ is the second end of e_{j_k} different from a_k, and as in [88] it follows that,

$$\beta_n u_{j_k,n}(a_{k'}) \longrightarrow 0 \quad \text{and} \quad \partial_x u_{j_k,n}(a_{k'}) \longrightarrow 0. \tag{4.45}$$

Now, let a_k be an external node of the tree $\mathcal{T}'$ obtained from $\mathcal{T}$ by removing all the strings attached to leaves. Using (4.45), the fifth condition in (4.7) and the continuity of $\underline{u}_n$ at a_k, we deduce that a_k satisfies (4.43). Then, by iterations, we have that for every j in J^s (even for $j = 1$)

$$-\frac{1}{2} \int_0^{\ell_j} \left(\left| \partial_x u_{j,n}(x) \right|^2 + \beta_n^2 \left| u_{j,n}(x) \right|^2 \right) dx \longrightarrow 0, \tag{4.46}$$

and

$$\beta_n u_{j,n}(a_{k'}) \longrightarrow 0 \text{ and } \partial_x u_{j,n}(a_{k'}) \longrightarrow 0, \tag{4.47}$$

where $a_{k'}$ is an end of e_j. In particular, using (4.40), we get

$$\left\| \partial_x u_{j,n} \right\|, \left\| v_{j,n} \right\| \longrightarrow 0.$$

If there is no beam in the tree, we conclude that $\|y_n\|_{\mathcal{H}} \to 0$ which contradicts the fact that $\|y_n\|_{\mathcal{H}} = 1$ and the proof is then complete.

Step 2 Now, we suppose that there is at least one beam. We denote by $\mathcal{G}$ the only maximal subgraph of beams of $\mathcal{T}$ and we will show that for every beam e_j in $\mathcal{G}$,

$$\left\| \partial_x^2 u_{j,n} \right\|, \left\| v_{j,n} \right\| \longrightarrow 0.$$

We start by beams ended by external nodes of $\mathcal{G}$ (i.e., those attached to stings). Let $a_k \in \mathcal{V}_{ext}(\mathcal{G})$, then

$$\beta_n u_{j_k,n}(a_k) \longrightarrow 0 \text{ and } \partial_x^3 u_{j_k,n}(a_k) \longrightarrow 0. \tag{4.48}$$

Indeed, if a_k is an external node of the initial tree $\mathcal{T}$, then (4.48) is due to (4.35) and (4.41), and damping conditions at external nodes, if a_k is attached to some strings, then it is due to (4.47) by taking into account the continuity condition of $\underline{u}_n$ and the fifth condition in (4.7) at inner nodes.

Let j be in J^b and q_j be a function in $C^2([0, \ell_j], \mathbb{C})$ such that $\partial_x^2 q_j = 0$. We want to calculate the real part of the inner product of (4.39) with $q_j \partial_x u_{j,n}$.

Straightforward calculations give

$$Re\left(\left\langle -\beta_n^2 u_{j,n}, q_j \partial_x u_{j,n}\right\rangle\right) + Re\left(\left\langle \partial_x^4 u_{j,n}, q_j \partial_x u_{j,n}\right\rangle\right) = -\frac{1}{2}\beta_n^2 \left|u_{j,n}(x)\right|^2 q_j(x)\Big|_0^{\ell_j}$$

$$+\frac{1}{2}\int_0^{\ell_j} \beta_n^2 \left|u_{j,n}\right|^2 \partial_x q_j dx + Re\left(\partial_x^3 u_{j,n}(x) q_j(x) \partial_x \overline{u}_{j,n}(x)\Big|_0^{\ell_j}\right)$$

$$-\frac{1}{2}\left|\partial_x^2 u_{j,n}(x)\right|^2 q_j(x)\Big|_0^{\ell_j} + \frac{3}{2}\int_0^{\ell_j}\left|\partial_x^2 u_{j,n}\right|^2 \partial_x q_j dx$$

$$-Re\left(\partial_x^2 u_{j,n}(x) \partial_x \overline{u_{j,n}} \partial_x q_j(x)\Big|_0^{\ell_j}\right),$$

and

$$Re\left(\langle g_{j,n} + i\beta_n f_{j,n}, q_j \partial_x u_{j,n}\rangle\right) = Re\left(\int_0^{\ell_j} g_{j,n}\partial_x \overline{u_{j,n}} q_j dx\right)$$

$$- Re\left(i\beta_n \int_0^{\ell_j} \partial_x(q_j f_{j,n})\overline{u}_{j,n} dx\right) + Re\, i\beta_n f_{j,n}(x)q_j(x)\overline{u}_{j,n}(x)\Big|_0^{\ell_j}.$$

Since $g_{j,n}$, $f_{j,n}$ and $\partial_x(q_j f_{j,n})$ converge to 0 in $L^2(0, \ell_j)$ and $i\beta_n u_{j,n}$ and $\partial_x u_{j,n}$ are bounded in $L^2(0, \ell_j)$, the first and the second terms of the right member of the previous equality converge to 0. It follows

$$-\frac{1}{2}\beta_n^2 \left|u_{j,n}(x)\right|^2 q_j(x)\Big|_0^{\ell_j} + Re\left(\partial_x^3 u_{j,n}(x) q_j(x) \partial_x \overline{u_{j,n}}(x)\Big|_0^{\ell_j}\right)$$

$$-\frac{1}{2}\left|\partial_x^2 u_{j,n}(x)\right|^2 q_j(x)\Big|_0^{\ell_j}$$

$$- Re\left(\partial_x^2 u_{j,n}(x) \partial_x \overline{u_{j,n}}(x) \partial_x q_j(x)\Big|_0^{\ell_j}\right) - Re\left(i\beta_n f_{j,n}(x)q_j(x)\overline{u_{j,n}}(x)\right)\Big|_0^{\ell_j}$$

$$+\frac{1}{2}\int_0^{\ell_j} \beta_n^2 \left|u_{j,n}\right|^2 \partial_x q_j dx + \frac{3}{2}\int_0^{\ell_j}\left|\partial_x^2 u_{j,n}\right|^2 \partial_x q_j dx \longrightarrow 0. \qquad (4.49)$$

In particular, if a_k is in $\mathcal{V}_{ext}(\mathcal{G})$, then with $j = j_k$ and $q_{j_k}(x) = x$ or $q_{j_k}(x) = \ell_{j_k} - x$ (4.49) becomes, using that $\partial_x u_{j_k,n}(a_k) = 0$,

$$-\frac{\ell_{j_k}}{2}\left|\partial_x^2 u_{j_k,n}(a_k)\right|^2 + Re\left(d_{k j_k}\partial_x^2 u_{j_k,n}(a_{k'})\partial_x \overline{u_{j_k,n}}(a_{k'})\right) + \frac{1}{2}\int_0^{\ell_{j_k}} \beta_n^2 \left|u_{j_k,n}\right|^2 dx$$

$$+\frac{3}{2}\int_0^{\ell_{j_k}}\left|\partial_x^2 u_{j_k,n}\right|^2 dx \to 0, \qquad (4.50)$$

where $a_{k'}$ is the end of e_{j_k} different from a_k.

Multiplying (4.39) by $\frac{1}{\beta_n^{1/2}} e^{-\beta_n^{1/2}(\ell_{jk}-x)}$ or by $\frac{1}{\beta_n^{1/2}} e^{-\beta_n^{1/2}x}$, then, as in [88], we obtain

$$\frac{1}{\beta_n^{1/2}} d_{kj_k} \partial_x^3 u_{j_k,n}(a_k) + \varepsilon \partial_x^2 u_{j_k,n}(a_k) + \varepsilon \beta_n u_{j_k,n}(a_k) \longrightarrow 0,$$

with $\varepsilon \in \{-1, 1\}$. Since $\partial_x^3 u_{j_k,n}(a_k)$ and $\beta_n u_{j_k,n}(a_k)$ tend to 0, we deduce that

$$\partial_x^2 u_{j_k,n}(a_k) \longrightarrow 0. \tag{4.51}$$

Hence, (4.50) can be rewritten as

$$Re\left(d_{kj_k} \partial_x^2 u_{j_k,n}(a_{k'}) \partial_x \overline{u_{j_k,n}}(a_{k'})\right) + \frac{1}{2} \int_0^{\ell_{jk}} \beta_n^2 \left|u_{j_k,n}\right|^2 dx$$

$$+ \frac{3}{2} \int_0^{\ell_{jk}} \left|\partial_x^2 u_{j_k,n}\right|^2 dx \longrightarrow 0. \tag{4.52}$$

Now, we rewrite (4.49), with $q_j = 1$ and $j = j_k$,

$$-\frac{1}{2} \beta_n^2 \left|u_{j_k,n}(a_{k'})\right|^2 + Re\left(\partial_x^3 u_{j_k,n}(a_{k'}) \partial_x \overline{u_{j_k,n}}(a_{k'})\right) - \frac{1}{2} \left|\partial_x^2 u_{j_k,n}(a_{k'})\right|^2$$

$$- Re\left(i\beta_n f_{j_k,n}(a_{k'}) \overline{u_{j_k,n}}(a_{k'})\right) \to 0. \tag{4.53}$$

For $j \in J_{k'}(\mathcal{G})$, multiplying (4.39) by $\frac{1}{\beta_n^{1/2}} e^{-\beta_n^{1/2}x}$ or by $\frac{1}{\beta_n^{1/2}} e^{-\beta_n^{1/2}(\ell_j-x)}$, we get

$$\frac{1}{\beta_n^{1/2}} d_{k'j} \partial_x^3 u_{j,n}(a_{k'}) + \varepsilon \partial_x^2 u_{j,n}(a_{k'}) + \beta_n^{1/2} d_{k'j} \partial_x u_{j,n}(a_{k'}) + \varepsilon \beta_n u_{j,n}(a_{k'}) \longrightarrow 0, \tag{4.54}$$

with $\varepsilon \in \{-1, 1\}$. In the case of $a_{k'} \neq a_1$, summing over $j \in J_{k'}(\mathcal{G})$, then by taking into account the continuity condition of $\underline{u}_n$ and $\partial_x^2 \underline{u}_n$, the fourth condition in (4.6), and the fifth in (4.7) at $a_{k'}$ we get

$$\partial_x^2 u_{j_k,n}(a_{k'}) + \beta_n u_{j_k,n}(a_{k'}) \longrightarrow 0.$$

It yields from (4.54),

$$\frac{1}{\beta_n^{1/2}} \partial_x^3 u_{j_k,n}(a_{k'}) + \beta_n^{1/2} \partial_x u_{j_k,n}(a_{k'}) := \alpha_{j_k,n} \longrightarrow 0, \tag{4.55}$$

which holds also if $a_{k'} = a_1$ (due to (4.54) and the boundary conditions at a_1). Hence, for any positive real number a, we have

$$
\begin{aligned}
Re\left(\partial_x^3 u_{j_k,n}(a_{k'})\partial_x \overline{u}_{j_k,n}(a_{k'})\right) &= Re\left(\frac{\partial_x^3 u_{j_k,n}(a_{k'})}{\beta_n^{1/2}}\beta_n^{1/2}\partial_x\overline{u}_{j_k,n}(a_{k'})\right) \\
&= Re\left((\alpha_{j_k,n} - \beta_n^{1/2}\partial_x u_{j_k,n}(a_{k'}))\beta_n^{1/2}\partial_x\overline{u}_{j_k,n}(a_{k'})\right) \\
&\le -\beta_n\left|\partial_x u_{j_k,n}(a_{k'})\right|^2 + \tfrac{a}{2}\beta_n\left|\partial_x u_{j_k,n}(a_{k'})\right|^2 \\
&\quad + \tfrac{1}{2a}\left|\alpha_{j_k,n}\right|^2 .
\end{aligned}
$$

$$(4.56)$$

Moreover, for any real positive number b we have

$$
-Re\left(\mathbf{i}\beta_n f_{j_k,n}(a_{k'})\overline{u}_{j_k,n}(a_{k'})\right) \le \frac{b}{2}\beta_n^2\left|u_{j_k,n}(a_{k'})\right|^2 + \frac{1}{2b}\left|f_{j_k,n}(a_{k'})\right|^2 . \qquad (4.57)
$$

Taking $j = j_k$ and combining (4.56), (4.57), and (4.53), we obtain the following framing for j_k:

$$
\begin{aligned}
&-\frac{1}{2}\beta_n^2\left|u_{j_k,n}(a_{k'})\right|^2 + Re\left(\partial_x^3 u_{j_k,n}(a_{k'})\partial_x\overline{u}_{j_k,n}(a_{k'})\right) - \frac{1}{2}\left|\partial_x^2 u_{j_k,n}(a_{k'})\right|^2 \\
&\quad - Re\left(\mathbf{i}\beta_n f_{j_k,n}(a_{k'})\overline{u}_{j_k,n}(a_{k'})\right) - \frac{1}{2b}\left|f_{j_k,n}(a_{k'})\right|^2 - \frac{1}{2a}\left|\alpha_{j_k,n}\right|^2 \\
&\le (-\frac{1}{2}+\frac{b}{2})\beta_n^2\left|u_{j_k,n}(a_{k'})\right|^2 - \frac{1}{2}\left|\partial_x^2 u_{j_k,n}(a_{k'})\right|^2 \\
&\quad + (-1+\frac{a}{2})\beta_n\left|\partial_x u_{j_k,n}(a_{k'})\right|^2 \le 0,
\end{aligned}
$$

with $a = b = \frac{1}{2}$. This implies, using that $f_{j_k,n}(a_{k'}) \longrightarrow 0$ and $\frac{1}{2a}\left|\alpha_{j,n}\right|^2 \longrightarrow 0$ the following properties:

$$
\beta_n u_{j_k,n}(a_{k'}) \longrightarrow 0, \quad \partial_x^2 u_{j_k,n}(a_{k'}) \longrightarrow 0, \quad \sqrt{\beta_n}\partial_x u_{j_k,n}(a_{k'}) \longrightarrow 0,
$$

$$
\text{and } \frac{1}{\sqrt{\beta_n}}\partial_x^3 u_{j_k,n}(a_{k'}) \longrightarrow 0. \qquad (4.58)
$$

We have used (4.55) for the last property. Moreover, $Re\left(\partial_x^2 u_{j_k,n}(a_{k'})\partial_x\overline{u}_{j_k,n}(a_{k'})\right)$ tends to 0 as n goes to infinity, then (4.52) leads to

$$
\frac{1}{2}\int_0^{\ell_{j_k}} \beta_n^2\left|u_{j_k,n}\right|^2 dx + \frac{3}{2}\int_0^{\ell_{j_k}} \left|\partial_x^2 u_{j_k,n}\right|^2 dx \longrightarrow 0.
$$

Now, let a_k be an external node of the graph $\mathcal{G}'$ obtained from $\mathcal{G}$ by removing all the edges ended by external nodes in $\mathcal{G}$. Using (4.58), the fifth condition in (4.7), and the continuity of $\underline{u}_n$ at a_k, we deduce that

$$\beta_n u_{j_k,n}(a_k) \longrightarrow 0, \quad \partial_x^2 u_{j_k,n}(a_k) \longrightarrow 0, \quad \sqrt{\beta_n} \partial_x u_{j_k,n}(a_k) \longrightarrow 0,$$

$$\text{and } \frac{1}{\sqrt{\beta_n}} \partial_x^3 u_{j_k,n}(a_k) \longrightarrow 0.$$

Then by iterations, we have that for every beam e_j (even if $j = 1$)

$$\frac{1}{2} \int_0^{\ell_j} \beta_n^2 \left| u_{j,n}(x) \right|^2 dx + \frac{3}{2} \int_0^{\ell_j} \left| \partial_x^2 u_{j,n}(x) \right|^2 dx \longrightarrow 0. \tag{4.59}$$

Finally, using (4.59), it yields

$$\left\| \partial_x^2 u_{j,n} \right\|, \beta_n \left\| u_{j,n} \right\| \longrightarrow 0.$$

In conclusion, $\|y_n\|_{\mathcal{H}}$ converge to 0, which contradicts the hypothesis that $\|y_n\|_{\mathcal{H}} = 1$. $\qquad\qquad\square$

2.3   Polynomial Stability

In this section we suppose that there is at least a beam following a string (Fig. 4.2). We will prove that the solution of the whole system $(\mathcal{S})$ is polynomially stable and non-exponentially stable at least in a simple case of two components.

Recall that r is the number of layers of beams in $\mathcal{T}$ not containing e_1. Then we have the following result:

Theorem 4.6 *If at least one beam follows a string from the root to a leaf, then the C_0-semigroup $(T(t)_{t \geq 0}$ is polynomially stable. More precisely: There are $C > 0$ such that*

$$\left\| e^{t\mathcal{A}} y_0 \right\|_{\mathcal{H}} \leq \frac{C}{t^{1/r}} \|y_0\|_{\mathcal{D}(\mathcal{A})}$$

for every $y_0 \in \mathcal{D}(\mathcal{A})$.

Proof We have proved that $i\mathbb{R} \subset \rho(\mathcal{A})$ (proof of Theorem 4.4) then, in view of Theorem 1.26, it suffices to prove that (1.12) holds for $\alpha = r$. Suppose the conclusion is false. Then there exists a sequence (β_n) of real numbers, without loss of generality, with $\beta_n \longrightarrow +\infty$, and a sequence of vectors $(y_n) = (\underline{u}_n, \underline{v}_n)$ in $\mathcal{D}(\mathcal{A})$ with $\|y_n\|_{\mathcal{H}} = 1$, such that

$$\left\| \beta_n^r (i\beta_n I - \mathcal{A}) y_n \right\|_{\mathcal{H}} \longrightarrow 0,$$

which is equivalent to

$$\beta_n^r(i\beta_n u_{j,n} - v_{j,n}) = f_{j,n} \longrightarrow 0, \quad \text{in } H^1(0, \ell_j), \quad j \text{ in } J^s, \tag{4.60}$$

$$\beta_n^r(i\beta_n u_{j,n} - v_{j,n}) = f_{j,n} \longrightarrow 0, \quad \text{in } H^2(0, \ell_j), \quad j \text{ in } J^b, \tag{4.61}$$

$$\beta_n^r(i\beta_n v_{j,n} - \partial_x^2 u_{j,n}) = g_{j,n} \longrightarrow 0, \quad \text{in } L^2(0, \ell_j), \quad j \text{ in } J^s, \tag{4.62}$$

$$\beta_n^r(i\beta_n v_{j,n} + \partial_x^4 u_{j,n}) = g_{j,n} \longrightarrow 0, \quad \text{in } L^2(0, \ell_j), \quad j \text{ in } J^b. \tag{4.63}$$

Then,

$$-\beta_n^r(\beta_n^2 u_{j,n} + \partial_x^2 u_{j,n}) = g_{j,n} + i\beta_n f_{j,n}, \quad j \text{ in } J^s, \tag{4.64}$$

$$\beta_n^r(-\beta_n^2 u_{j,n} + \partial_x^4 u_{j,n}) = g_{j,n} + i\beta_n f_{j,n}, \quad j \text{ in } J^b, \tag{4.65}$$

and

$$\beta_n^r \left\| v_{j,n} \right\|^2 - \beta_n^{2+r} \left\| u_{j,n} \right\|^2 \longrightarrow 0, \quad j = 1, \dots, N. \tag{4.66}$$

Since $Re(\langle \beta_n^r(i\beta_n - A)y_n, y_n \rangle_{\mathcal{H}}) = \sum_{a_k \in \mathcal{V}_{ext}^*} \beta_n^r \left| v_{j_k,n}(a_k) \right|^2$, we obtain

$$\beta_n^{\frac{r}{2}} \left| v_{j_k,n}(a_k) \right| \longrightarrow 0, \quad \text{for } a_k \in \mathcal{V}_{ext}^*.$$

We will prove that $\|y_n\|_{\mathcal{H}}$ converge to zero, which contradicts the fact that $\|y_n\|_{\mathcal{H}} = 1$.

The rest of the proof will be decomposed into four steps, in which we need several times the real part of the inner product of (4.64) and (4.65) with $q_j \partial_x u_{j,n}$ respectively, where $q_j = 1$ or $q_j = x$ or $q_j = \ell_j - x$. They give

$$\frac{1}{2}\beta_n^{2+r} \left| u_{j,n}(x) \right|^2 q_j(x) \Big|_0^{\ell_j} + \frac{1}{2}\beta_n^r \left| \partial_x u_{j,n}(x) \right|^2 q_j(x) \Big|_0^{\ell_j}$$

$$-\frac{1}{2}\beta_n^r \int_0^{\ell_j} \left(\left| \partial_x u_{j,n}(x) \right|^2 + \beta_n^2 \left| u_{j,n}(x) \right|^2 \right) \partial_x q_j(x) dx$$

$$+ Re\left(i\beta_n f_{j,n}(x) q_j(x) \overline{u}_{j,n}(x) \right)\Big|_0^{\ell_j} \longrightarrow 0, \tag{4.67}$$

in the first case (that is $j \in J^s$) and

$$-\frac{1}{2}\beta_n^{2+r} \left| u_{j,n}(x) \right|^2 q_j(x) \Big|_0^{\ell_j} + Re\left(\beta_n^r \partial_x^3 u_{j,n}(x) q_j(x) \partial_x \overline{u}_{j,n}(x) \Big|_0^{\ell_j} \right)$$

$$-\frac{1}{2}\beta_n^r \left| \partial_x^2 u_{j,n}(x) \right|^2 q_j(x) \Big|_0^{\ell_j}$$

$$- Re\left(\beta_n^r \partial_x^2 u_{j,n}(x) \partial_x \overline{u_{j,n}} \partial_x q_j(x)\Big|_0^{\ell_j}\right) - Re\left(i\beta_n f_{j,n}(x) q_j(x)\overline{u_{j,n}}(x)\right)\Big|_0^{\ell_j}$$

$$+ \frac{1}{2}\beta_n^r \int_0^{\ell_j} \beta_n^2 |u_{j,n}|^2 \partial_x q_j dx + \frac{3}{2}\beta_n^r \int_0^{\ell_j} \left|\partial_x^2 u_{j,n}\right|^2 \partial_x q_j dx \longrightarrow 0, \qquad (4.68)$$

in the second case (that is $j \in J^b$).

Step 1 We will prove that for every edge e_j of the first layer of strings we have

$$\beta_n^{1+\frac{r}{2}} \|u_{j,n}\|, \beta_n^{\frac{r}{2}} \|\partial_x u_{j,n}\| \longrightarrow 0. \qquad (4.69)$$

Let $\mathcal{G}$ be a maximal subgraph of NFB-strings of $\mathcal{T}$ (i.e., a maximal subgraph in the first layer of $\mathcal{T}$).

Let $a_k \in \mathcal{V}_{ext}^s$. By using (4.60) and damping conditions at external nodes, we have

$$\beta_n^{1+\frac{r}{2}} u_{j_k,n}(a_k) \longrightarrow 0 \text{ and } \beta_n^{\frac{r}{2}} \partial_x u_{j_k,n}(a_k) \longrightarrow 0. \qquad (4.70)$$

Then, from (4.70) and (4.67), with $j = j_k$ and $q_{j_k}(x) = x$ or $q_{j_k}(x) = \ell_{j_k} - x$, we have that (4.69) holds for $j = j_k$.

Moreover, as in the previous theorem, (4.69) holds for any edge e_j of $\mathcal{G}$, and if a_k is an end of such edge then

$$\beta_n^{1+\frac{r}{2}} u_{j,n}(a_k) \longrightarrow 0 \text{ and } \beta_n^{\frac{r}{2}} \partial_x u_{j,n}(a_k) \longrightarrow 0. \qquad (4.71)$$

In fact, (4.69) and (4.67) hold for every string e_j in the first layer of strings.

Step 2 Now, we assume that we have removed the first layer of strings. Let $\mathcal{G}$ be a subgraph of beams, maximal for the property that every edge is NFS ($\mathcal{G}$ is a maximal subgraph in the second layer of $\mathcal{T}$). Note that there is no string attached to a node in $\mathcal{V}_{ext}(\mathcal{G}) \cup \mathcal{V}_{int}(\mathcal{G})$.

Let $a_k \in \mathcal{V}_{ext}(\mathcal{G})$. If a_k is an external node of $\mathcal{T}$, then by (4.60) and damping conditions at external nodes, we get

$$\beta_n^{1+\frac{r}{2}} u_{j_k,n}(a_k) \longrightarrow 0, \ \beta_n^{\frac{r}{2}} \partial_x^3 u_{j_k,n}(a_k) \longrightarrow 0 \text{ and } \partial_x u_{j_k,n}(a_k) = 0,$$

which holds if a_k is not an external node of $\mathcal{T}$, due to the continuity condition of u_n, the fourth condition in (4.6), and the fifth in (4.7) at a_k and (4.71).

Again, as in the previous proof, using (4.68) in some iterations, we have that for every beam e_j of $\mathcal{G}$, not attached to a string,

$$\beta_n^{1+\frac{r}{2}} \|u_{j,n}\|, \ \beta_n^{\frac{r}{2}} \|\partial_x^2 u_{j,n}\| \longrightarrow 0, \qquad (4.72)$$

and if a_k is a node of $\mathcal{G}$, not belonging to a string, then for every $j \in J_k(\mathcal{G})$,

$$\beta_n^{1+\frac{r}{2}} u_{j,n}(a_k) \longrightarrow 0, \quad \beta_n^{\frac{r}{2}-\frac{1}{2}} \partial_x^3 u_{j,n}(a_k) \longrightarrow 0, \quad \beta_n^{\frac{r}{2}+\frac{1}{2}} \partial_x u_{j,n}(a_k) \to 0,$$

and $\beta_n^{\frac{r}{2}} \partial_x^2 u_{j,n}(a_k) \longrightarrow 0.$ \hfill (4.73)

Moreover, we will prove that (4.73) holds even if a_k is an end of a string. Note that, in this case, a_k must be the outer node of $\mathcal{G}$ and that every beam attached to a_k lies in $\mathcal{G}$.

Taking $q_j = 1$ in (4.68), with j in J_k^b, we obtain

$$-\frac{1}{2}\beta_n^{2+r} |u_{j,n}(a_k)|^2 + Re\left(\beta_n^r \partial_x^3 u_{j,n}(a_k) \partial_x \overline{u}_{j,n}(a_k)\right) - \frac{1}{2}\beta_n^r \left|\partial_x^2 u_{j,n}(a_k)\right|^2 \to 0.$$
\hfill (4.74)

Multiplying (4.65) by $\frac{1}{\beta_n} e^{-\beta_n^{1/2} x}$ or by $\frac{1}{\beta_n} e^{-\beta_n^{1/2}(\ell_j - x)}$ leads to

$$\beta_n^{r-\frac{1}{2}} d_{kj} \partial_x^3 u_{j,n}(a_k) + \varepsilon \beta_n^r \partial_x^2 u_{j,n}(a_k) + \beta_n^{r+\frac{1}{2}} d_{kj} \partial_x u_{j,n}(a_k) + \varepsilon \beta_n^{1+r} u_{j,n}(a_k) \longrightarrow 0.$$
\hfill (4.75)

Summing over j in J_k^b we get, using the fifth condition in (4.7) and fourth in (4.6),

$$\beta_n^{r-\frac{1}{2}} \sum_{j \in J_k^s} d_{kj} \partial_x u_{j,n}(a_k) + R\varepsilon \left(\beta_n^r \partial_x^2 u_{j,n}(a_k) + \beta_n^{1+r} u_{j,n}(a_k)\right) \longrightarrow 0,$$

where R is the cardinal of J_k^b. Returning back to (4.75),

$$\beta_n^{r-\frac{1}{2}} d_{kj} \partial_x^3 u_{j,n}(a_k) - \frac{1}{R}\beta_n^{r-\frac{1}{2}} \sum_{i \in J_k^s} d_{ki} \partial_x u_{i,n}(a_k) + \beta_n^{r+\frac{1}{2}} d_{kj} \partial_x u_{j,n}(a_k) := \gamma_{j,n} \longrightarrow 0.$$
\hfill (4.76)

Multiplying now (4.76) by $\beta_n^{\frac{1}{2}} d_{kj} \partial_x u_{j,n}(a_k)$ at the left, and summing the real parts, we obtain

$$\beta_n^r \sum_{j \in J_k^b} Re\left(\partial_x u_{j,n}(a_k) \partial_x^3 \overline{u}_{j,n}(a_k)\right) + \beta_n^{r+1} \sum_{j \in J_k^b} \left|\partial_x u_{j,n}(a_k)\right|^2$$

$$= \sum_{j \in J_k^b} Re\left(\overline{\gamma}_{j,n} \beta_n^{\frac{1}{2}} d_{kj} \partial_x u_{j,n}(a_k)\right),$$
\hfill (4.77)

due to condition fifth in (4.7). We deduce that

$$\beta_n^r \sum_{j \in J_k^b} Re \left(\partial_x u_{j,n}(a_k) \partial_x^3 \overline{u_{j,n}}(a_k) \right)$$

$$\leq -\beta_n^{r+1} \sum_{j \in J_k^b} \left| \partial_x u_{j,n}(a_k) \right|^2 + \frac{1}{2} \beta_n^{r+1} \sum_{j \in J_k^b} \left| \partial_x u_{j,n}(a_k) \right|^2$$

$$+ \frac{1}{2\beta_n^r} \sum_{j \in J_k^b} \left| \gamma_{j,n} \right|^2. \tag{4.78}$$

Therefore, we can deduce from (4.77), after summing over j in J_k^b, that (4.73) holds for all j in J_k^b. Moreover, by taking $q_j = x$ or $q_j = \ell_j - x$ in (4.68), we obtain that (4.72) holds for every j in J_k^b. Thus, (4.72) and (4.73) are verified for every beam e_j of the second layer of $\mathcal{T}$.

Step 3 Let $\mathcal{G}$ be a maximal subgraph of NFB-strings in the new tree obtained by removing the last layer of beams.

For every node a_k in $\mathcal{V}_{ext}(\mathcal{G})$ we have

$$\beta_n^{1+\frac{r}{2}} u_{j_k,n}(a_k) \longrightarrow 0 \text{ and } \beta_n^{\frac{r-1}{2}} \partial_x u_{j_k,n}(a_k) \longrightarrow 0,$$

due to the transmission conditions at a_k and (4.73). As in *Step 1*, if e_j is a string of $\mathcal{G}$ then

$$\beta_n^{1+\frac{r-1}{2}} \left\| u_{j,n} \right\|, \beta_n^{\frac{r-1}{2}} \left\| \partial_x u_{j,n} \right\| \longrightarrow 0 \tag{4.79}$$

and

$$\beta_n^{1+\frac{r-1}{2}} u_{j,n}(a_{k'}) \longrightarrow 0 \text{ and } \beta_n^{\frac{r-1}{2}} \partial_x u_{j,n}(a_{k'}) \longrightarrow 0,$$

where $a_{k'}$ is an end of e_j.

If $e_1 \notin \mathcal{G}$. Suppose that we remove the last layer of strings and let $\mathcal{G}'$ be a maximal subgraph of NFS-beams in the new tree. Then
$\forall a_k \in \mathcal{V}_{ext}(\mathcal{G}')$

$$\beta_n^{1+\frac{r-1}{2}} u_{j_k,n}(a_k) \longrightarrow 0, \quad \beta_n^{\frac{r-1}{2}} \partial_x^3 u_{j_k,n}(a_k) \longrightarrow 0 \text{ and } \beta_n^{\frac{r}{2}+\frac{1}{2}} \partial_x u_{j_k,n}(a_k) \longrightarrow 0.$$

As in *Step 2* , we have by iteration, that for every beam e_j of $\mathcal{G}'$ (even if $j = 1$),

$$\beta_n^{1+\frac{r-1}{2}} \left\| u_{j,n} \right\|, \ \beta_n^{\frac{r-1}{2}} \left\| \partial_x^2 u_{j,n} \right\| \longrightarrow 0, \tag{4.80}$$

and for every node a_k of e_j,

$$\beta_n^{1+\frac{r-1}{2}} u_{j,n}(a_k) \longrightarrow 0, \ \ \beta_n^{\frac{r-1}{2}-\frac{1}{2}} \partial_x^3 u_{j,n}(a_k) \longrightarrow 0, \ \ \beta_n^{\frac{r-1}{2}+\frac{1}{2}} \partial_x u_{j,n}(a_k) \to 0$$

and $\beta_n^{\frac{r-1}{2}} \partial_x^2 u_{j,n}(a_k) \longrightarrow 0.$

Step 4 We iterate such procedure in $\mathcal{T} - \{e_1\}$ (from leaves to the root).

If e_1 is a string, we have, using the transmission and damping conditions at a_2,

$$\beta_n^{1+\frac{r-r}{2}} u_{1,n}(a_2) \longrightarrow 0 \text{ and } \beta_n^{\frac{r-r}{2}} \partial_x u_{1,n}(a_2) \longrightarrow 0,$$

then, by taking $j = 1$ in (4.67), we get

$$\beta_n^{1+\frac{r-r}{2}} \left\| u_{1,n} \right\|, \ \beta_n^{\frac{r-r}{2}} \left\| \partial_x^2 u_{1,n} \right\| \longrightarrow 0. \tag{4.81}$$

If e_1 is a beam, then

$$\beta_n^{1+\frac{r-r}{2}} u_{1,n}(a_2) \longrightarrow 0, \ \ \beta_n^{\frac{r-r}{2}-\frac{1}{2}} \partial_x^3 u_{1,n}(a_2) \longrightarrow 0 \text{ and } \beta_n^{\frac{r-r}{2}+\frac{1}{2}} \partial_x u_{1,n}(a_2) \to 0. \tag{4.82}$$

Moreover,

$$\beta_n^{\frac{r-r}{2}} \partial_x^2 u_{1,n}(a_2) \longrightarrow 0.$$

Using continuity conditions of $\underline{u}_n$ and $\partial_x^2 u_n$, and the fourth condition in (4.6) and fifth in (4.7) at a_2 , (4.68) leads to

$$\beta_n^{1+\frac{r-r}{2}} \left\| u_{1,n} \right\|, \ \beta_n^{\frac{r-r}{2}} \left\| \partial_x^2 u_{1,n} \right\| \longrightarrow 0. \tag{4.83}$$

In conclusion $\|y_n\|$ converge to 0, which contradicts the hypothesis that $\|y_n\|_{\mathcal{H}} = 1$. $\qquad\qquad\qquad\square$

2.4 Lack of Exponential Stability

Now, we consider a reduced system composed of one string e_1 and one beam e_2 such that $\ell_1 = \ell_2 = \pi$ and with control is applied on the beam. Precisely, we consider the system

$$(S_0) : \begin{cases} u_{1,tt} - u_{1,xx} = 0 \text{ in } (0, \pi) \times (0, \infty), \\ u_{2,tt} + u_{2,xxxx} = 0 \text{ in } (0, \pi) \times (0, \infty), \\ u_1(0, t) = u_2(0, t), \ u_{2,x}(0, t) = 0, \ u_{2,xxx}(0, t) = u_{1,x}(0, t), \\ u_1(\pi, t) = 0, \ u_{2,xxx}(\pi, t) = u_{2,t}(\pi, t), \ u_{2,x}(\pi, t) = 0, \\ u_j(x, 0) = u_j^0(x), \ u_{j,t}(x, 0) = u_j^1(x), \ j = 1, 2. \end{cases}$$

In view of Theorem 4.6, the system (S_0) is polynomial stable, and we will prove that it is not exponentially stable. Note that if the control is applied on the string instead of the beam, then the system is exponentially stable (by Theorem 4.5).

Theorem 4.7 *The system (S_0) is not exponentially stable in the energy space $\mathcal{H}$.*

Proof We prove that the corresponding semigroup $(T(t))_{t \geq 0}$ is not exponentially stable.

For $n \in \mathbb{N}$, such that $\sqrt{n}$ is integer and even let $\beta_n = n^2 + 2\sqrt{n} + \frac{1}{n}$ and $f_n = (0, 0, -\sin \beta_n x, 0)$, then $\beta_n \to +\infty$ and f_n is in $\mathcal{H}$ and is bounded. Let $y_n = (u_{1,n}, u_{2,n}, v_{1,n}, v_{2,n}) \in \mathcal{D}(\mathcal{A})$ such that $(\mathcal{A} - i\beta_n)y_n = f_n$. We will prove that $y_n \to +\infty$.

We have

$$\beta_n^2 u_{1,n} + \partial_x^2 u_{1,n} = \sin \beta_n x, \tag{4.84}$$

$$-\beta_n^2 u_{2,n} + \partial_x^4 u_{2,n} = 0. \tag{4.85}$$

Then $u_{1,n}$ and $u_{2,n}$ are of the form

$$u_{1,n} = c_1 \sin(\beta_n x) + \left(-\frac{x}{2\beta_n} + c_2\right) \cos(\beta_n x),$$

$$u_{2,n} = d_1 \sin(\sqrt{\beta_n} x) + d_2 \cos(\sqrt{\beta_n} x) + d_3 \sinh(\sqrt{\beta_n} x) + d_4 \cosh(\sqrt{\beta_n} x).$$

The transmission and boundary conditions are rewritten as follows:

$$d_2 + d_4 = c_2, \tag{4.86}$$

$$\sqrt{\beta_n}(d_1 + d_3) = 0, \tag{4.87}$$

$$\beta_n^{3/2}(-d_1 + d_3) = -\frac{1}{2\beta_n} + \beta_n c_1, \tag{4.88}$$

and

$$c_1 \sin(\beta_n \pi) + \left(-\frac{\pi}{2\beta_n} + c_2\right) \cos(\beta_n \pi) = 0,$$

(4.89)

$$d_1 \cos(\sqrt{\beta_n}\pi) - d_2 \sin(\sqrt{\beta_n}\pi) + d_3 \cosh(\sqrt{\beta_n}\pi) + d_4 \sinh(\sqrt{\beta_n}\pi) = 0,$$

(4.90)

and

$$\beta_n^{3/2}(-d_1 \cos(\sqrt{\beta_n}\pi) + d_2 \sin(\sqrt{\beta_n}\pi) + d_3 \cosh(\sqrt{\beta_n}\pi) + d_4 \sinh(\sqrt{\beta_n}\pi))$$
$$= \mathbf{i}\beta_n (d_1 \sin(\sqrt{\beta_n}\pi) + d_2 \cos(\sqrt{\beta_n}\pi) + d_3 \sinh(\sqrt{\beta_n}\pi)$$
$$+ d_4 \cosh(\sqrt{\beta_n}\pi)).$$

(4.91)

Summing (4.90) and (4.91), we obtain

$$2\sqrt{\beta_n}(d_3 \cosh(\sqrt{\beta_n}\pi) + d_4 \sinh(\sqrt{\beta_n}\pi))$$
$$= \mathbf{i}(d_1 \sin(\sqrt{\beta_n}\pi) + d_2 \cos(\sqrt{\beta_n}\pi) + d_3 \sinh(\sqrt{\beta_n}\pi)$$
$$+ d_4 \cosh(\sqrt{\beta_n}\pi)).$$

(4.92)

Substituting (4.86) and (4.87) into (4.90) leads to

$$d_4 = \frac{\cosh(\sqrt{\beta_n}\pi) - \cos(\sqrt{\beta_n}\pi)}{\sinh(\sqrt{\beta_n}\pi) + \sin(\sqrt{\beta_n}\pi)} d_1 + \frac{\sin(\sqrt{\beta_n}\pi)}{\sinh(\sqrt{\beta_n}\pi) + \sin(\sqrt{\beta_n}\pi)} c_2.$$

(4.93)

Now, by substituting (4.86)–(4.88) and (4.93) into (4.92), we get

$$2d_1 \left(\sqrt{\beta_n}h(\sqrt{\beta_n}\pi) - 2\beta_n \tan(\beta_n\pi)\sin(\sqrt{\beta_n}\pi)\sinh(\sqrt{\beta_n}\pi)\right.$$
$$\left. + \mathbf{i}\left(1 - \cosh(\sqrt{\beta_n}\pi)\cos(\sqrt{\beta_n}\pi) + \sqrt{\beta_n}\tan(\beta_n\pi)h(\sqrt{\beta_n}\pi)\right)\right)$$
$$= \left(2\sqrt{\beta_n}\sin(\sqrt{\beta_n}\pi)\sinh(\sqrt{\beta_n}\pi) - \mathbf{i}h(\sqrt{\beta_n}\pi)\right)$$
$$\times \left(-\frac{1}{2\beta_n^2}\tan(\beta_n\pi) + \frac{\pi}{2\beta_n}\right),$$

(4.94)

with

$$h(\sqrt{\beta_n}\pi) = \cosh(\sqrt{\beta_n}\pi)\sin(\sqrt{\beta_n}\pi) + \sinh(\sqrt{\beta_n}\pi)\cos(\sqrt{\beta_n}\pi).$$

We want to prove that $\beta_n^{3/2}d_1$ is equivalent to $\sqrt{n}\frac{\pi^2}{2}$ as n goes to infinity.

Since $\beta_n = n^2 + 2\sqrt{n} + \frac{1}{n}$, then $\sqrt{\beta_n} = n + \frac{1}{\sqrt{n}} = n(1 + o(\frac{1}{\sqrt{n}}))$ and

$$\sin(\sqrt{\beta_n}\pi) = \sin(\frac{\pi}{\sqrt{n}}) = \frac{\pi}{\sqrt{n}} + o(\frac{1}{\sqrt{n}}),$$

$$\cos(\sqrt{\beta_n}\pi) = \cos(\frac{\pi}{\sqrt{n}}) = 1 + o(1),$$

$$\tan(\beta_n\pi) = \tan(\frac{\pi}{n}) = \frac{\pi}{n} + o(\frac{1}{n}),$$

$$\cosh(\sqrt{\beta_n}\pi) = \cosh((n + \frac{1}{\sqrt{n}})\pi) = \frac{e^{n\pi}}{2}(1 + o(1)),$$

$$\sinh(\sqrt{\beta_n}\pi) = \frac{e^{n\pi}}{2}(1 + o(1)),$$

then

$$h(\sqrt{\beta_n}\pi) = \frac{e^{n\pi}}{2}(1 + o(1)),$$

$$\sqrt{\beta_n} \sinh(\sqrt{\beta_n}\pi) \sin(\sqrt{\beta_n}\pi) \tan(\beta_n\pi) = \frac{e^{n\pi}}{2}o(1),$$

$$\sqrt{\beta_n} \sinh(\sqrt{\beta_n}\pi) \sin(\sqrt{\beta_n}\pi) = \pi\sqrt{n}\frac{e^{n\pi}}{2}(1 + o(1)).$$

Hence, (4.94) implies

$$2d_1\sqrt{\beta_n}\frac{e^{n\pi}}{2}(1 + o(1)) = \frac{e^{n\pi}}{2}\sqrt{n}(2\pi + o(1))\left(-\frac{1}{2\beta_n^2}O(\frac{1}{n}) + \frac{\pi}{2\beta_n}\right)$$

that leads to

$$2\beta_n^{3/2}d_1 \sim \pi^2\sqrt{n}. \tag{4.95}$$

when n tends to infinity.

Now taking the real part of the inner product of (4.84) with $(\pi - x)\partial_x u_{1,n}$, we get

$$-\frac{\pi}{2}\left|-\frac{1}{2\beta_n} + \beta_n c_1\right|^2 - \frac{\pi}{2}|\beta_n c_2|^2 = -\frac{1}{2}\left(\beta_n^2 \|u_{1,n}\|^2 + \|\partial_x u_{1,n}\|^2\right)$$

$$+ Re\left(\int_0^\pi \sin(\beta_n x)(\pi - x)\partial_x\overline{u_{1,n}}dx\right).$$

Then, by taking into account (4.87-4.88) and (4.95), $\beta_n^2 \|u_{1,n}\|^2 + \|\partial_x u_{1,n}\|^2$ must be not bounded. In conclusion y_n is not bounded. $\quad\square$

Remark 4.8 In view of Theorem 1.26, a slight modification in the proof of the previous theorem allows us to deduce that the semigroup associated to system $(\mathcal{S}_0)$ cannot be polynomially stable of order α for every $\alpha > 2$. Moreover, such result persists if we consider two components of any length.

Remark 4.9 We believe that this last result of non-exponential stability remains true in the general case, when at least a beam following a string (Fig. 4.2).

3 Comments

3.1 Comment 1

As in [37], the Authors take in [36] a transmission problem coupling heat and wave equations on a star-shaped network. For a simple model (Fig. 4.3) one can see [107] and [106] where some results of controllability and polynomial stability are obtained.

Now, let us describe briefly the star-shaped network $\mathcal{G}$ (Fig. 4.4) considered in [36]. The edges (segments) e_j, $j = 1, \ldots, N$ occupy the intervals $(0, \ell_j)$, $\ell_j > 0$, respectively. The common node is identified to $x = 0$. The heat equations arise on the intervals $(0, \ell_k)$, $k = 1, 2, \ldots, N_1$ $(0 < N_1 < N)$ in the network with state θ_k, respectively; the wave equations hold on the intervals $(0, \ell_j)$, $k = N_1 + 1, N_1 +$

Fig. 4.3 Coupled heat-wave

Wave   Heat

Fig. 4.4 Star-shaped
network: heat equations (red),
wave equations (black)

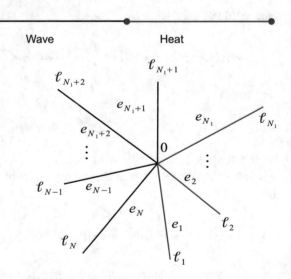

$2, \ldots, N$ in the network with state $(u_j, u_{j,t})$. The authors choose the following heat-wave system on star-shaped network.

$$
\begin{cases}
\theta_{k,t}(x,t) - \theta_{k,xx}(x,t) = 0, & x \in (0, \ell_j), \ k = 1, 2, \cdots, N_1, \ t > 0, \\
u_{j,tt}(x,t) - u_{j,xx}(x,t) = 0, & x \in (0, \ell_j), \ j = N_1 + 1, \cdots, \cdots, N, \ t > 0, \\
\theta_k(\ell_k, t) = u_j(\ell_j, t), & \forall \, j, k = 1, 2, \cdots, N, \ t > 0, \\
u_j(\ell_j, t) = 0, & j = 1, 2, \cdots, N, \ t > 0, \\
u_j(0, t) = u_k(0, t), & \forall \, j, k = 1, 2, \cdots, N, \ t > 0, \\
\theta_k(\ell_k, t) = 0, & k = 1, 2, \cdots, N_1, \ t > 0, \\
\theta_k(0, t) = u_j(0, t), & \forall \, k = 1, 2, \cdots, N_1, \ j = N_1 + 1, N_1 + 2, \cdots, N, \ t > 0, \\
\displaystyle\sum_{j=N_1+1}^{N} u_{j,x}(0, t) + \sum_{k=1}^{N_1} \theta_{k,x}(0, t) = 0, \ t > 0, \\
\theta^k(t = 0) = \theta_k^0, \ k = 1, 2, \cdots, N_1, \\
u_j(t = 0) = u_j^0, \ u_{j,t}(t = 0) = u_j^1, \ j = N_1 + 1, N_1 + 2, \cdots, N,
\end{cases}
$$

$$(4.96)$$

where $\left(\left(\theta_k^0 \right)_{k=1}^{N_1}, \left(u_j^0 \right)_{j=N_1+1}^{N}, \left(u_j^1 \right)_{j=N_1+1}^{N} \right)$ is the given initial state.

The energy of this system is defined as follows:

$$
E(t) = \frac{1}{2} \sum_{k=1}^{N_1} \int_0^{\ell_k} |\theta_k|^2 dx + \frac{1}{2} \sum_{j=N_1 1}^{N} \int_0^{\ell_j} \left(|u_{j,x}|^2 + |u_{j,t}|^2 \right) dx.
$$

It satisfies

$$
\frac{dE(t)}{dt} = - \sum_{k=1}^{N_1} \int_0^{\ell_k} |\theta_{k,t}|^2 dx \le 0
$$

and therefore the energy is decreasing.

Then, the authors wrote the system as an abstract Cauchy problem in some appropriate Hilbert space $\mathcal{H}$:

$$
\begin{cases}
\dfrac{d}{dt} y = \mathcal{A} y, \\
y(0) = y^0,
\end{cases}
$$

where $y = (\underline{u}, \underline{u}_t, \underline{\theta})$ and $y^0 = (\underline{u}^0, \underline{u}^1, \underline{\theta}^0) \in \mathcal{H}$ is given.

The authors proved the following results.

Theorem 4.10 *The energy of the system (4.96) decays to zero as $t \to \infty$ if and only if one of the following two conditions is fulfilled:*

(1) $N - N_1 = 1$,
(2) $N - N_1 \geq 2$ *and* $\ell_i / \ell_j \notin \mathbb{Q}$, $i, j = N_1 + 1, N_1 + 2, \cdots, N$, $i \neq j$.

Theorem 4.11 *The energy of system (4.96) does not decay exponentially as $t \to \infty$, as soon as the network involves at least one wave equation.*

The authors examine more deeply the case $N - N_1 > 1$, in particular, they give the following estimate for $N - N_1 = 1$.

Theorem 4.12 *Case $N - N_1 = 1$: The heat-wave star contains only one edge described by a wave equation. The energy of the corresponding system decays polynomially as follows:*
   There exists a positive constant C such that

$$E(t) \leq Ct^{-4} \|(\underline{u}^0, \underline{u}^1, \underline{\theta}^0)\|^2_{\mathcal{D}(\mathcal{A})}, \ \forall t \geq 0$$

for every $(\underline{u}^0, \underline{u}^1, \underline{\theta}^0) \in \mathcal{D}(\mathcal{A})$. This decay rate is sharp.

3.2   Comment 2

Some of the following are recent works related to the classical and fractional heat equation on graphs: [63, 67, 72, 90, 99].

Chapter 5
Feedback Stabilization of a Simplified Model of Fluid–Structure Interaction on a Tree

In this chapter, we study the stability of a model of fluid propagating in a 1-d network, under some feedbacks applied at exterior nodes, with the presence of point mass at inner nodes, see Fig. 5.1. At rest, the network coincides with the tree $\mathcal{T} = (E, \mathcal{V})$, where $E = \{e_1, \ldots, e_N\}$ is the set of edges and $\mathcal{V} = \{a_1, \ldots, a_{N+1}\}$ is the set of nodes (vertices). We assume that a_1 is the root of $\mathcal{T}$, which will be denoted by $\mathcal{R}$, that e_1 is the edge containing $\mathcal{R}$, and that a_2 is its vertex different from $\mathcal{R}$.

More precisely, we consider the following initial and boundary value problem:

$$\begin{cases} y_{j,tt} - y_{j,xx} = 0 \text{ in } (0, \ell_j) \times (0, \infty), \quad j \in J := \{1, \ldots, N\}, \\[2mm] \displaystyle\sum_{j \in J_k} d_{kj} y_{j,x}(a_k, t) = s'_k(t), \quad k \in I_{int}, \\ s''_k(t) + s_k(t) = -y_t(a_k, t), \quad k \in I_{int}, \\ y_j(a_k, t) = y_l(a_k, t), \quad j, l \in J_k, \quad k \in I_{int}, \\ y_1(a_1, t) = 0, \\ d_{kj_k} y_{j_k,x}(a_k, t) = -y_t(a_k, t), \quad k \in I^*_{ext}, \\[2mm] s_k(0) = s^0_k, \quad s'_k(0) = s^1_k, \quad k \in I_{int}, \\ y_j(x, 0) = y^0_j(x), \quad y_{j,t}(x, 0) = y^1_j(x), \quad x \in (0, \ell_j), \quad j \in J, \end{cases} \qquad (5.1)$$

where $y_j : [0, \ell_j] \times (0, \infty) \longrightarrow \mathbb{R}$, $j \in J$, represents the velocity potential of the fluid on the edge e_j and $s_k : (0, \infty) \longrightarrow \mathbb{R}$, $k \in I_{int}$, denotes the movement of the point mass occupant the node a_k. These functions allow us to identify the network with its rest graph.

This simplified model of fluid–structure interaction draws on the work of Ervedoza and Vanninathan [24]; they considered a fluid occupying a domain in two dimensions and a solid immersed in it and proved some results of controllability of such system, see also [92] for more details. Then, we consider a corresponding one-dimensional model (where the dimension of space is reduced to 1), with many

© The Author(s), under exclusive license to Springer Nature Switzerland AG 2022
K. Ammari, F. Shel, *Stability of Elastic Multi-Link Structures*, SpringerBriefs in Mathematics, https://doi.org/10.1007/978-3-030-86351-7_5

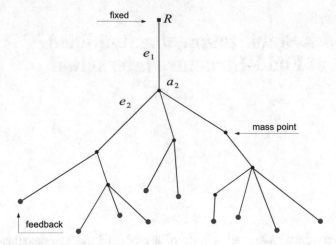

Fig. 5.1 A tree-shaped network

mass points (which model the structure). We refer to [17] where the authors gave some mathematical models of vibrations of fluid–solid structures corresponding to some physical situations, as the tube bundles vibrating inside a moving fluid in a nuclear reactor. The problem of fluid–structure interaction in one dimension has been studied by several authors. In [95], the authors studied the asymptotic behavior of a one-dimensional model of mass point moving in a fluid. They considered the same system in [96], but with a finite number of mass points floating in the fluid. Recently Tucsnak et al. [59] studied the controllability of a similar system. In [97], the authors modeled a biological problem (an intracranial saccular aneurysm), into a coupled fluid–structure interaction problem, in one dimension, consisting of a wave equation with a dynamical condition at one end.

Note that the pointwise (or boundary) stabilization on the wave equation has been treated during the last few years, see, for example, [9] for one string and [4, 7, 10] for some networks of strings.

The main result of this chapter asserts that, under some conditions, the energy of the solutions of the dissipative system decays exponentially to zero when the time tends to infinity. The method is based on a frequency domain method and a special analysis for the resolvent.

If $(y, \underline{s}) = ((y_j)_{j \in J}, (s_k)_{k \in I_{int}})$ is a solution of (5.1), we define the energy of $(y, \underline{s})$ at instant t by

$$E(t) = \frac{1}{2} \sum_{j \in J} \int_0^{\ell_j} \left(\left| y_{j,t}(x, t) \right|^2 + \left| y_{j,x}(x, t) \right|^2 \right) dx + \frac{1}{2} \sum_{k \in I_{int}} \left(\left| s_k'(t) \right|^2 + \left| s_k(t) \right|^2 \right).$$

Simple formal calculations show that a sufficiently smooth solution of (5.1) satisfies the energy estimate

$$E(0) - E(t) = \sum_{k \in I_{ext}^*} \int_0^t |y_{j_k,t}(a_k, s)|^2 \, ds, \quad \forall t \geq 0. \tag{5.2}$$

In particular, (5.2) implies that

$$E(t) \leq E(0), \quad \forall t \geq 0.$$

Define

$$V = \left\{ \underline{\Phi} \in \prod_{j \in J} H^1(0, \ell_j), \; \Phi_1(\mathcal{R}) = 0, \; \Phi_j(a_k) = \Phi_l(a_k), \; j, l \in J_k, \; k \in I_{int} \right\},$$

and the natural well-posedness space for (5.1) is

$$H = V \times \prod_{j \in J} L^2(0, \ell_j) \times \left(\prod_{k \in I_{int}} \mathbb{C} \right)^2$$

endowed with the inner product

$$\left\langle Z, \tilde{Z} \right\rangle_H = \sum_{j \in J} \left(\int_0^{\ell_j} \partial_x f_j(x) \partial_x \overline{\tilde{f}_j}(x) dx + \int_0^{\ell_j} g_j(x) \overline{\tilde{g}_j}(x) dx \right)$$
$$+ \sum_{k \in I_{int}} \left(c_k \overline{\tilde{c}_k} + d_k \overline{\tilde{d}_k} \right),$$

where $Z = (\underline{f}, \underline{g}, \underline{c}, \underline{d})$ and $\tilde{Z} = (\underline{\tilde{f}}, \underline{\tilde{g}}, \underline{\tilde{c}}, \underline{\tilde{d}})$.

Then, we can rewrite the system (5.1) as a first-order differential equation, by putting $z(t) = \left(\underline{y}(t), \underline{y}'(t), \underline{s}(t), \underline{s}'(t) \right)$:

$$z'(t) = \mathcal{A}z(t), \; z(0) = z^0 = \left(\underline{y}^0, \underline{y}^1, \underline{s}^0, \underline{s}^1 \right),$$

where

$$\mathcal{A}(\underline{y}, \underline{v}, \underline{p}, \underline{q}) = (\underline{v}, \partial_x^2 \underline{y}, \underline{q}, -\underline{p} - \underline{v}_\mathcal{M}), \; \forall (\underline{y}, \underline{v}, \underline{p}, \underline{q}) \in \mathcal{D}(\mathcal{A}),$$

with $\underline{v}_{\mathcal{M}} = (\underline{v}(a_k), k \in I_{int})$, and

$$\mathcal{D}(\mathcal{A}) = \left\{ (\underline{y}, \underline{v}, \underline{p}, \underline{q}) \in \left[\prod_{j \in J} H^2(0, \ell_j) \cap V \right] \times V \times \left(\prod_{k \in I_{int}} \mathbb{C} \right)^2 \text{ satisfying (5.3)} \right\},$$

where

$$\begin{cases} \sum_{j \in J_k} d_{kj} \partial_x y_j(a_k) = q_k, & \forall k \in I_{int}, \\ d_{kj} \partial_x y_{j_k}(a_k) = -v_{j_k}(a_k), & \forall k \in I^*_{ext}. \end{cases} \tag{5.3}$$

To simplify the notations, sometimes, we will write $\underline{y}(a_k)$ instead of $y_j(a_k)$ for $\underline{y}$ in V.

The outline of this work is the following. In Sect. 1, we prove the existence and uniqueness of solutions for system (5.1). Section 2 is devoted to proving the exponential stability of the associated semigroup. Finally, in Sect. 3, we prove the lack of exponential stability of a graph containing a circuit and a tree with two uncontrolled exterior nodes (Figs. 5.2 and 5.3). The Sect. 4 is devoted to the study of the case of a chain with non-equal mass points (Fig. 5.4).

Fig. 5.2 Circuit

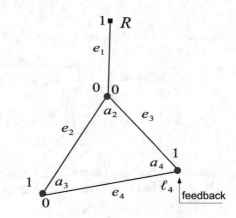

Fig. 5.3 Star

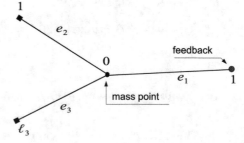

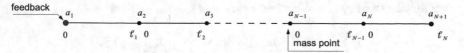

Fig. 5.4 A chain of strings

1 Well-Posedness

Lemma 5.1 *The operator $\mathcal{A}$ generates a C_0-semigroup of contraction $(S(t))_{t\geq 0}$ on H.*

Proof It is clear that the operator $\mathcal{A}$ is dissipative, and moreover, for every $z = (\underline{y}, \underline{v}, \underline{p}, \underline{q}) \in \mathcal{D}(\mathcal{A})$,

$$Re\left(\langle \mathcal{A}z, z\rangle_H\right) = -\sum_{k\in I_{ext}^*} \left|v_{j_k}(a_k)\right|^2 \leq 0. \tag{5.4}$$

Now, we prove that every positive real number λ belongs to $\rho(\mathcal{A})$, the resolvent set of $\mathcal{A}$. For this, let $Z = (\underline{f}, \underline{g}, \underline{c}, \underline{d}) \in H$, and we solve the equation

$$(\lambda - \mathcal{A})z = Z \tag{5.5}$$

with $z = (\underline{y}, \underline{v}, \underline{p}, \underline{q})$ in $\mathcal{D}(\mathcal{A})$.
We rewrite (5.5) explicitly

$$\begin{cases} \lambda y_j - v_j = f_j, & j \in J, \\ \lambda v_j - \partial_x^2 y_j = g_j, & j \in J, \\ \lambda p_k - q_k = c_k, & k \in I_{int}, \\ \lambda q_k + p_k + \underline{v}(a_k) = d_k, & k \in I_{int}. \end{cases} \tag{5.6}$$

We eliminate $(\underline{v}, \underline{q})$ in (5.6) to get

$$\lambda^2 y_j - \partial_x^2 y_j = g_j + \lambda f_j, \quad j \in J, \tag{5.7}$$

$$(\lambda^2 + 1)p_k + \underline{v}(a_k) = d_k + \lambda c_k, \quad k \in I_{int}. \tag{5.8}$$

Let $\underline{w}$ in V. Multiplying (5.7) by w_j in $L^2(0, \ell_j)$ and summing over $j \in J$,

$$\sum_{j\in J}\left(\lambda^2 \int_0^{\ell_j} y_j\overline{w_j}dx + \int_0^{\ell_j}\partial_x y_j\partial_x\overline{w_j}dx\right) - \sum_{k\in I_{int}}\overline{w}(a_k)q_k$$

$$+ \sum_{k\in I_{ext}^*}\overline{w}(a_k)\underline{v}(a_k) = \sum_{j\in J}\int_0^{\ell_j}(g_j + \lambda f_j)\overline{w_j}dx. \tag{5.9}$$

Multiplying (5.8) by $r_k \in \mathbb{C}$ and summing over $k \in I_{int}$, we get

$$(\lambda^2 + 1) \sum_{k \in I_{int}} p_k \overline{r_k} + \sum_{k \in I_{int}} \underline{v}(a_k) \overline{r_k} = \sum_{k \in I_{int}} (d_k + \lambda c_k) \overline{r_k}. \tag{5.10}$$

Summing (5.9) and (5.10), we get

$$\sum_{j \in J} \left(\lambda^2 \langle y_j, w_j \rangle + \langle \partial_x y_j, \partial_x w_j \rangle \right) + (\lambda^2 + 1) \sum_{k \in I_{int}} p_k \overline{r_k}$$

$$+ \sum_{k \in I_{int}} \left(\underline{v}(a_k) \overline{r_k} - \overline{\underline{w}(a_k)} q_k \right)$$

$$+ \sum_{k \in I^*_{ext}} \overline{\underline{w}(a_k)} \underline{v}(a_k) = \sum_{j \in J} \langle g_j + \lambda f_j, w_j \rangle + \sum_{k \in I_{int}} (d_k + \lambda c_k) \overline{r_k}$$

to obtain

$$a((\underline{y}, \underline{p}), (\underline{w}, \underline{r})) = F(\underline{w}, \underline{r}),$$

where

$$a((\underline{y}, \underline{p}), (\underline{w}, \underline{r})) = \sum_{j \in J} \left(\lambda^2 \langle y_j, w_j \rangle + \langle \partial_x y_j, \partial_x w_j \rangle \right) + (\lambda^2 + 1) \sum_{k \in I_{int}} p_k \overline{r_k}$$

$$+ \lambda \sum_{k \in I_{int}} \left(\underline{y}(a_k) \overline{r_k} - \overline{\underline{w}(a_k)} p_k \right) + \lambda \sum_{k \in I^*_{ext}} \overline{\underline{w}(a_k)} \underline{y}(a_k)$$

and

$$f(\underline{w}, \underline{r}) = \sum_{j \in J} \langle g_j + \lambda f_j, w_j \rangle + \sum_{k \in I_{int}} (d_k + \lambda c_k) \overline{r_k} - \sum_{k \in I_{int}} \overline{\underline{w}(a_k)} c_k$$

$$+ \sum_{k \in I_{int}} \underline{f}(a_k) \overline{r_k} + \sum_{k \in I^*_{ext}} \overline{\underline{w}(a_k)} \underline{f}(a_k).$$

a is a continuous sesquilinear form on $V \times \prod\limits_{k \in I_{int}} \mathbb{C}$ and F is a continuous anti-linear form on $V \times \prod\limits_{k \in I_{int}} \mathbb{C}$. Moreover,

$$a((\underline{w}, \underline{r}), (\underline{w}, \underline{r})) = \sum_{j \in J} \left(\lambda^2 \|w_j\|^2 + \|\partial_x w_j\|^2 \right) + (\lambda^2 + 1) \sum_{k \in I_{int}} |r_k|^2$$

$$- 2i\lambda Im \left(\sum_{k \in I_{int}} \overline{\underline{w}(a_k)} r_k \right) + \lambda \sum_{k \in I^*_{ext}} |\underline{w}(a_k)|^2.$$

We have

$$
\left| a((\underline{w}, \underline{r}), (\underline{w}, \underline{r})) \right| \geq \left[\sum_{j \in J} \left(\lambda^2 \|w_j\|^2 + \|\partial_x w_j\|^2 \right) + \lambda^2 \sum_{k \in I_{int}} |r_k|^2 \right],
$$

that is, a is coercive. The conclusion results immediately from the Lax–Milgram lemma.   □

According to Theorem 1.15, we have the following results:

Proposition 5.2 *Suppose that* $\left(\underline{y}^0, \underline{y}^1, \underline{s}^0, \underline{s}^1 \right) \in H$. *Then, the problem (5.1) admits a unique solution*

$$
(\underline{y}, \underline{y}', \underline{s}, \underline{s}') \in \mathcal{C}([0, +\infty); H).
$$

If $\left(\underline{y}^0, \underline{y}^1, \underline{s}^0, \underline{s}^1 \right) \in \mathcal{D}(\mathcal{A})$, *then*

$$
(\underline{y}, \underline{y}', \underline{s}, \underline{s}') \in \mathcal{C}([0, +\infty), \mathcal{D}(\mathcal{A})) \cap \mathcal{C}^1([0, +\infty); H).
$$

Moreover, $(\underline{y}, \underline{s})$ *satisfies the energy estimate (5.2).*

2 Exponential Stability

It is clear that if $\mathcal{T}$ contains an edge e_j, not attached to a leaf, with length $\ell_j \in \pi \mathbb{N}$, then $\mathbf{i}$ is eigenvalue of $\mathcal{A}$ with eigenvector $\underline{z} = (y, v, p, q)$ such that $y_j = \mathbf{i} \sin x$, $v_j = -\sin x$, $p_k = 1$, and $q_k = \mathbf{i}$ when a_k is the nearest end of e_j to the root $\mathcal{R}$, and all the other components of z are null.

In the following, the tree $\mathcal{T}$ is said to be a **Pi-tree** if the length of every edge not attached to a leaf is different from $m\pi$ for every m in $\mathbb{N}^*$. Then, we have the following result.

Lemma 5.3 *The spectrum of $\mathcal{A}$ contains no point on the imaginary axis if and only if $\mathcal{T}$ is a Pi-tree.*

Proof By the Sobolev embedding theorem, we can deduce that $(\mathcal{I} - \mathcal{A})^{-1}$ is a compact operator. Then, the spectrum of $\mathcal{A}$ only consists of eigenvalues. We will show that the equation

$$
\mathcal{A}z = \mathbf{i}\beta z \tag{5.11}
$$

with $z = (y, v, p, q) \in \mathcal{D}(\mathcal{A})$ and $\beta \in \mathbb{R}$ has only trivial solution.

By taking the inner product of (5.11) with $z \in H$ and using that

$$Re\left(\langle \mathcal{A}z, z \rangle_H\right) = - \sum_{k \in I^*_{ext}} \left| \underline{v}(a_k) \right|^2,$$

we obtain that

$$\underline{v}(a_k) = 0 \text{ for } k \in I^*_{ext}. \tag{5.12}$$

Furthermore, by the second condition in (5.3), we also have

$$\partial_x y(a_k) = 0 \text{ for } k \in I^*_{ext}. \tag{5.13}$$

Now, Eq. (5.11) can be rewritten explicitly as

$$v_j = i\beta y_j, \quad j \in J, \tag{5.14}$$

$$\partial_x^2 y_j = i\beta v_j, \quad j \in J, \tag{5.15}$$

$$q_k = i\beta p_k, \quad k \in I_{int}, \tag{5.16}$$

$$-p_k - \underline{v}(a_k) = i\beta q_\kappa, \quad k \in I_{int}. \tag{5.17}$$

If $\beta = 0$, then $\underline{v} = 0$, $q = 0$, and $p = 0$.
Multiplying the second equation in the above system by y_j and then summing over j, we obtain

$$\sum_{j \in J} \left\| \partial_x y_j \right\|^2 = 0,$$

which implies, using the continuity condition of y at inner nodes and its Dirichlet condition at $\mathcal{R}$, that $y = 0$.
   Next, we suppose that $\beta \neq 0$. From (5.12) and (5.14), and we get

$$y_{j_k}(a_k) = 0 \text{ for } k \in I^*_{ext}.$$

Using (5.14)–(5.17), we have

$$\beta^2 y_j + \partial_x^2 y_j = 0, \quad j \in J, \tag{5.18}$$

$$(\beta^2 - 1)p_k = \underline{v}(a_k), \quad k \in I_{int}. \tag{5.19}$$

The function y_j, $j \in J$, is then of the form

$$y_j = \alpha_1 \sin(\beta x) + \alpha_2 \cos(\beta x).$$

Using (5.12), (5.13), and (5.18), we obtain that $y_{j_k} = 0$, for every k in I_{ext}^*, then using (5.14), $v_{j_k} = 0$, for every k in I_{ext}^*.

For the sequel of the proof, we consider two cases.

First Case: $\beta^2 = 1$ From (5.19), (5.14), and the Dirichlet condition at $\mathcal{R}$, we deduce that $y_j(0) = y_j(\ell_j) = 0$ for $j \in J$. Since $\mathcal{T}$ is a Pi-tree, we conclude that $y_j = 0$ for $j \in J$. Return back to the balance conditions and (5.16), one can deduce that $q = p = 0$ and hence $z = 0$.

Second Case: $\beta^2 \neq 1$ Let a_k be the second end of an edge e_j attached to a leaf. Since y_j and v_j are zero, and using (5.19) and (5.16), we deduce that $p_k = 0$ and $q_k = 0$. Next, let e_l be an internal edge (i.e., not containing a leaf) attached at a node a_k to an edge e_j ended by a leaf. Using again the balance condition and the continuity of y at a_k, we obtain that $y_l(a_k) = 0$ and $\partial_x y_l(a_k) = 0$. Then, by (5.18) , $y_l = 0$. We iterate such procedure from leaves to root to conclude that $z = 0$.

$\square$

The main result of this chapter concerns the precise asymptotic behavior of the solution of (5.1). Our technique is based on a frequency domain method and a special analysis for the resolvent.

Recall that the system (5.1) is said to be exponentially stable if there exist two constants $M, \omega > 0$, such that for all $\left(y^0, y^1, s^0, s^1\right) \in H$,

$$E(t) \leq M e^{-\omega t} \left\| \left(y^0, y^1, s^0, s^1\right) \right\|_H^2, \quad \forall t \geq 0.$$

Then, our main result is the following:

Theorem 5.4 *The system defined by Eq. (5.1) is exponentially stable if and only if $\mathcal{T}$ is a Pi-tree.*

Proof The system defined by Eq. (5.1) is exponentially stable if and only if the C_0-semigroup of contraction $(S(t))_{t\geq 0}$, generated by $\mathcal{A}$, is exponentially stable.

By Theorem 1.25, it suffices to show that the operator $\mathcal{A}$ satisfies the following two conditions:

$$\rho(\mathcal{A}) \supset \{i\beta \mid \beta \in \mathbb{R}\} \equiv i\mathbb{R} \quad \text{and} \quad \limsup_{\beta \in \mathbb{R}, |\beta| \to \infty} \left\| (i\beta - \mathcal{A})^{-1} \right\| < \infty. \quad (5.20)$$

By Lemma 5.3, the first condition in (5.20) is satisfied if and only if $\mathcal{T}$ is a Pi-tree. Then, we suppose that $\mathcal{T}$ is a Pi-tree, and by contradiction, we suppose that the second condition in (5.20) is false. Then, there exist a sequence of real numbers $\beta_n \to \infty$ ($\beta_n > 0$ without loss of generality) and a sequence of vector $z_n = (y_n, v_n, p_n, q_n) \in \mathcal{D}(\mathcal{A})$ with $\|z_n\|_H = 1$ such that

$$\|(i\beta_n I - \mathcal{A})z_n\|_H \longrightarrow 0 \quad \text{as} \quad n \longrightarrow \infty, \quad (5.21)$$

i.e.,

$$i\beta_n y_{j,n} - v_{j,n} \equiv f_{j,n} \longrightarrow 0 \quad \text{in} \quad H^1(0, \ell_j), \qquad (5.22)$$

$$i\beta_n v_{j,n} - \partial_x^2 y_{j,n} \equiv g_{j,n} \longrightarrow 0 \quad \text{in} \quad L^2(0, \ell_j), \qquad (5.23)$$

$$i\beta_n p_{k,n} - q_{k,n} \equiv h_{k,n} \longrightarrow 0 \quad \text{in} \quad \mathbb{C}, \qquad (5.24)$$

$$i\beta_n q_{k,n} + p_{k,n} + \underline{v}_n(a_k) \equiv r_{k,n} \longrightarrow 0 \quad \text{in} \quad \mathbb{C}, \qquad (5.25)$$

for every j in J and every k in I_{int}. Our goal is to derive from (5.21) that $\|z_n\|_H$ converge to zero, thus, a contradiction.

The proof is divided into three steps.

First Step Recall that for every j in J,

$$\frac{1}{\beta_n^{1/2}} \|v_j\|_\infty \leq \frac{C_1}{\beta_n^{1/2}} \|\partial_x v_j\|^{1/2} \|v_j\|^{1/2} + \frac{C_2}{\beta_n^{1/2}} \|v_j\|,$$

for some positives constants C_1 and C_2 (we have used the Gagliardo–Nirenberg inequality (1.14)). This implies, using (5.22), that $\frac{\|v_{j,n}\|_\infty}{\beta_n^{1/2}}$ is bounded.

Then, for every k in I_{int}, by dividing (5.25) by β_n, we deduce that $q_{k,n}$ converge to zero, and then $\beta_n p_{k,n}$ converge to zero in view of (5.24). In particular, $p_{k,n}$ converge to zero.

Second Step Notice that from (5.4), we have

$$\|(i\beta_n I - \mathcal{A})z_n\|_H \geq \left| Re \langle i\beta_n I - \mathcal{A})z_n, z_n \rangle_H \right| = \sum_{k \in I_{ext}^*} \left| \underline{v}_n(a_k) \right|^2. \qquad (5.26)$$

Then, by (5.21),

$$\left| \underline{v}_n(a_k) \right| \longrightarrow 0, \quad \forall k \in I_{ext}^*.$$

This further leads to

$$\left| \beta_n \underline{y}_n(a_k) \right| \longrightarrow 0, \quad \forall k \in I_{ext}^* \qquad (5.27)$$

due to (5.22) and the trace theorem.

We also have

$$\frac{dy_{j_k}}{dx}(a_k) \longrightarrow 0, \quad \forall k \in I_{ext}^*. \qquad (5.28)$$

Third Step Substitute (5.22) into (5.23) to get

$$- \beta_n^2 y_{j,n} - \partial_x^2 y_{j,n} = g_{j,n} + \mathbf{i}\beta_n f_{j,n}, \quad j \in J. \tag{5.29}$$

Next, we take the inner product of (5.29) with $\partial_x y_{j,n} b$ in $L^2(0, \ell_j)$ for $b \in C^1[0, \ell_j]$, we get

$$\frac{1}{2}\beta_n^2 \left[|y_{j,n}(x)|^2 b(x) \right]_0^{\ell_j} + \frac{1}{2} \left[|\partial_x y_{j,n}(x)|^2 b(x) \right]_0^{\ell_j}$$

$$+ \left[Re \left(\mathbf{i}\beta_n f_{j,n}(x)\overline{y_{j,n}}(x)b(x) \right) \right]_0^{\ell_j}$$

$$- \frac{1}{2} \int_0^{\ell_j} \left(|\beta_n y_{j,n}|^2 + |\partial_x y_{j,n}|^2 \right) \partial_x b(x) dx \longrightarrow 0.$$

Using (5.27) and (5.28), this leads to

$$\frac{1}{2} \int_0^{\ell_{jk}} \left(|\beta_n y_{jk,n}|^2 + |\partial_x y_{jk,n}|^2 \right) dx \longrightarrow 0$$

for every k in I_{ext}^*, by taking $b = x$ or $b = \ell_{jk} - x$. It follows that

$$\frac{1}{2}\beta_n^2 |y_{jk,n}(a_s)|^2 + \frac{1}{2} |\partial_x y_{jk,n}(a_s)|^2 + Re \left(\mathbf{i}\beta_n f_{jk,n}(a_s)\overline{y_{jk,n}}(a_s) \right) \longrightarrow 0, \tag{5.30}$$

where a_s is the end of e_{jk}, different from a_k.

We will show that all the terms in the first members of (5.30) converge to zero. To do that, we use the following inequality:

$$Re \left(\mathbf{i}\beta_n f_{jk,n}(a_s)\overline{y_{jk,n}}(a_s) \right) \geq -\frac{1}{4}\beta_n^2 |y_{jk,n}(a_s)|^2 - |f_{jk,n}(a_s)|^2$$

to obtain

$$\frac{1}{2}\beta_n^2 |y_{jk,n}(a_s)|^2 + \frac{1}{2} |\partial_x y_{jk,n}(a_s)|^2 + Re \left(\mathbf{i}\beta_n f_{jk,n}(a_s)\overline{y_{jk,n}}(a_s) \right) + |f_{jk,n}(a_s)|^2$$

$$\geq \frac{1}{4}\beta_n^2 |y_{jk,n}(a_s)|^2 + \frac{1}{2} |\partial_x y_{jk,n}(a_s)|^2 \geq 0. \tag{5.31}$$

In addition, we have $f_{jk,n}(a_s) \longrightarrow 0$. Then, (5.31) combined with (5.30) implies that

$$\beta_n y_{jk,n}(a_s) \longrightarrow 0, \text{ and } \partial_x y_{jk,n}(a_s) \longrightarrow 0.$$

Furthermore, we also have

$$Re\left(i\beta_n f_{j_k,n}(a_s)\overline{y_{j_k,n}}(a_s)\right) \longrightarrow 0.$$

We then conclude by iteration, as in [88], that for every j in J,

$$\int_0^{\ell_j} \left(\left|\beta_n y_{j,n}\right|^2 + \left|\partial_x y_{j,n}\right|^2\right) dx \longrightarrow 0.$$

Finally, in view of (5.22), we also get

$$\|v_{j,n}\| \longrightarrow 0, \quad \text{for } j \in J.$$

In conclusion, $p_{k,n}$ and $q_{k,n}$ converge to zero for every k in I_{int}, and $\|\partial_x y_{j,n}\|$ and $\|v_{j,n}\|$ converge to zero for every j in J, which implies that $\|z_n\|_H \longrightarrow 0$ clearly contradicts (5.21). $\qquad\square$

3   Two Examples of Non-exponential Stability

In this section, we consider two particular cases. In the first case (Fig. 5.2), there is a circuit in the graph. We prove that, even with much more controls, the exponential stability fails. In the second case (Fig. 5.3), we eliminate a control of a leaf. We prove that the new system is also not exponentially stable.

3.1   A Circuit (Fig. 5.2)

In this part, we suppose that $\mathcal{T}$ contains a circuit (Fig. 5.2), with $\ell_1 = \ell_2 = \ell_3 = 1$ and with feedbacks at each inner node. Then, the second equation in (5.1) will be

$$\sum_{j\in J_k} d_{kj}\, y_{j,x}(a_k, t) = s'_1(t) - \underline{y}_t(a_k, t), \quad a_k \in I_{int},$$

and we can rewrite the system (5.1) in the Hilbert space H as

$$z'(t) = \mathcal{A}z(t).$$

The associated state space is

$$H = V \times \left(\left(L^2(0,1)\right)^3 \times L^2(0, \ell_4)\right) \times \left(\mathbb{C}^3\right)^2$$

with

$$V = \left\{ \underline{\Phi} \in \left(H^1(0,1) \right)^3 \times H^1(0, \ell_4), \ \Phi_1(1) = 0, \ \Phi_1(0) = \Phi_2(0) = \Phi_3(0), \right.$$
$$\left. \Phi_2(1) = \Phi_4(0), \ \Phi_3(1) = \Phi_4(\ell_4) \right\}.$$

The system (5.1) can be rewritten in the Hilbert space H as

$$z'(t) = \mathcal{A}z(t), \quad z(0) = \left(\underline{y}^0, \underline{y}^1, \underline{s}^0, \underline{s}^1 \right),$$

where $\mathcal{A}$ is the operator defined on H by

$$\mathcal{A}(\underline{y}, \underline{v}, \underline{p}, \underline{q}) = (\underline{v}, \partial_x^2 \underline{y}, \underline{q}, -\underline{p} - \underline{v}_{\mathcal{M}}), \ \forall (\underline{y}, \underline{v}, \underline{p}, \underline{q}) \in \mathcal{D}(\mathcal{A}),$$

with $\underline{v}_{\mathcal{M}} = (\underline{v}(a_2), \underline{v}(a_3), \underline{v}(a_4))$ and

$$\mathcal{D}(\mathcal{A}) = \left\{ (\underline{y}, \underline{v}, \underline{p}, \underline{q}) \in \left[\left(\left(H^2(0,1) \right)^3 \times H^2(0, \ell_4) \right) \cap V \right] \times V \times \left(\mathbb{C}^3 \right)^2 \right.$$

$$\left. \text{satisfying (5.32)} \right\},$$

where

$$\begin{cases} -\sum_{j=1}^3 \partial_x y_j(0) = q_2 - \underline{v}(a_2), \\ \partial_x y_2(1) - \partial_x y_4(0) = q_3 - \underline{v}(a_3), \\ \partial_x y_3(1) + \partial_x y_4(\ell_4) = q_4 - \underline{v}(a_4). \end{cases} \tag{5.32}$$

The operator $\mathcal{A}$ generates a C_0-semigroup of contraction $(S(t))_{t \geq 0}$ satisfying the first result of asymptotic behavior.

Theorem 5.5 $(S(t))_{t \geq 0}$ *is asymptotically stable if and only if ℓ_4 is irrational and not in $\pi \mathbb{Z}$.*

Proof First, if ℓ_4 is in $\pi \mathbb{Z}$, then $\mathbf{i}$ is an eigenvalue of $\mathcal{A}$ (as in the case of a tree), and if $\ell_4 = \frac{a}{b}$ with a and b integer, then $\mathbf{i}b\pi$ is an eigenvalue of $\mathcal{A}$.

Now, we suppose that $\ell_4 \notin \mathbb{Q}$ and $\ell_4 \notin \pi \mathbb{Z}$. We only need to prove that $\mathbf{i}\mathbb{R} \subset \rho(\mathcal{A})$. For this, we will prove that the equation

$$\mathcal{A}z = \mathbf{i}\beta z \tag{5.33}$$

with $z = (\underline{y}, \underline{v}, \underline{p}, \underline{q}) \in \mathcal{D}(\mathcal{A})$ and $\beta \in \mathbb{R}$ has only trivial solution.

The real part of the inner product of (5.33) with $z \in H$ is

$$Re\left(\langle \mathcal{A}z, z \rangle_H\right) = -\sum_{k \in I_{int}} \left|\underline{v}(a_k)\right|^2 = -\left(\left|\underline{v}(a_2)\right|^2 + \left|\underline{v}(a_3)\right|^2 + \left|\underline{v}(a_4)\right|^2\right); \quad (5.34)$$

then $v(a_k) = 0$ for every $k \in 2, 3, 4$. Now Eq. (5.33) can be rewritten as follows:

$$v_j = \mathbf{i}\beta y_j, \quad j \in J, \quad (5.35)$$

$$\partial_x^2 y_j = \mathbf{i}\beta v_j, \quad j \in J, \quad (5.36)$$

$$q_k = \mathbf{i}\beta p_k, \quad k \in 2, 3, 4, \quad (5.37)$$

$$-p_k - \underline{v}(a_k) = \mathbf{i}\beta q_\kappa, \quad k \in 2, 3, 4. \quad (5.38)$$

If $\beta = 0$, then, as in for the initial example, we show that $\underline{y} = 0$.

Next, we suppose that $\beta \neq 0$. We have

$$\underline{y}(a_k) = 0 \text{ for every } k \in 2, 3, 4 \quad (5.39)$$

in view of (5.34)–(5.36), and

$$\beta^2 y_j + \partial_x^2 y_j = 0, \quad j \in J, \quad (5.40)$$

$$(\beta^2 - 1)p_k = 0, \quad k \in 2, 3, 4. \quad (5.41)$$

By using (5.35)–(5.38), the condition (5.39) implies that the solution of (5.40) is of the form

$$y_j = \alpha_j \sin(\beta x),$$

where $\alpha_j \in \mathbb{R}$. As in the case of a tree, we consider two cases: $\beta^2 = 1$ and $\beta^2 \neq 1$.

First Case: $\beta^2 = 1$ We have, using again (5.36), $\alpha_1 = \alpha_2 = \alpha_3 = 0$, and since $\ell_4 \notin \pi\mathbb{Z}$, $\alpha_4 = 0$. Return back to the balance condition and (5.37), at inner nodes, one can deduce that $\underline{q} = \underline{p} = 0$ and hence $z = 0$.

Second Case: $\beta^2 \neq 1$ We have $p_k = q_k = 0$ for every k in I_{int}. Now, if $\beta \notin \pi\mathbb{Z}$, then (5.38) gives that $\alpha_1 = \alpha_2 = \alpha_3 = 0$, and the balance condition at a_3 implies that $\alpha_4 = 0$. If $\beta \in \pi\mathbb{Z}$, then (5.38) gives $\alpha_4 = 0$, since $\ell_4 \notin \mathbb{Q}$. Using again the balance condition, respectively, at a_3, a_4, and a_2, we obtain that $\alpha_2 = \alpha_3 = \alpha_1 = 0$. Then, $z = 0$.

$\square$

Theorem 5.6 *The semigroup $(S(t))_{t \geq 0}$ is not exponentially stable, even if ℓ_4 is irrational and not in $\pi\mathbb{Z}$.*

Proof To prove that $(S(t))_{t \geq 0}$ is not exponentially stable, we consider the sequence f_n of vectors of H defined by $f_n = (0, \underline{g}_n, 0, 0)$, where $\underline{g}_n = (0, -\sin \beta_n x, 0, 0)$ and β_n is a sequence of real numbers satisfying $\beta_n \longmapsto +\infty$ and that will be defined later. We then prove that the sequence $z_n = (\underline{y}_n, \underline{v}_n, \underline{p}_n, \underline{q}_n)$ of elements of $\mathcal{D}(\mathcal{A})$ such that

$$(i\beta_n - \mathcal{A})z_n = f_n$$

is not bounded.

The sequences $\underline{y}_n$ and $\underline{q}_n$ should satisfy

$$\begin{cases} \beta_n^2 y_{j,n} + \partial_x^2 y_{j,n} = 0, & \text{for } j = 1, 3, 4, \\ \beta_n^2 y_{2,n} + \partial_x^2 y_{2,n} = \sin \beta_n x, \\ q_{k,n} = -\frac{\beta_n^2}{\beta_n^2 - 1} \underline{y}_n(a_k) = -\beta_n c_n \underline{y}_n(a_k), & \text{for } k = 2, 3, 4, \end{cases}$$

with $c_n = \frac{\beta_n}{\beta_n^2 - 1}$. Then, for $j = 1, 3, 4$, there exist two complex numbers a_j and b_j (depending of n) such that

$$\begin{cases} y_{j,n} = a_j \sin(\beta_n x) + b_j \cos(\beta_n x), \\ \partial_x y_{j,n} = -\beta_n b_j \sin(\beta_n x) + \beta_n a_j \cos(\beta_n x), \end{cases}$$

and there exist two complex numbers a_2 and b_2 (depending of n) such that

$$\begin{cases} y_{2,n} = a_2 \sin(\beta_n x) + (-\frac{x}{2\beta_n} + b_2) \cos(\beta_n x), \\ \partial_x y_{2,n} = (\frac{x}{2} - \beta_n b_2) \sin(\beta_n x) + (-\frac{1}{2\beta_n} + \beta_n a_2) \cos(\beta_n x). \end{cases}$$

The boundary and transmission conditions are expressed as follows:

$$\begin{cases} a_1 \sin(\beta_n) + b_1 \cos(\beta_n) = 0, \\ b_1 = b_2 = b_3, \\ -\frac{1}{2\beta_n} + \beta_n a_1 + \beta_n a_2 + \beta_n a_3 = (i + c_n)\beta_n b_1, \\ a_2 \sin(\beta_n) + (-\frac{1}{2\beta_n} + b_2) \cos(\beta_n) = b_4, \\ \beta_n a_4 - \left((\frac{1}{2} - \beta_n b_2) \sin(\beta_n) + (-\frac{1}{2\beta_n} + \beta_n a_2) \cos(\beta_n) \right) = (i + c_n)\beta_n b_4, \\ a_3 \sin(\beta_n) + b_3 \cos(\beta_n) = a_4 \sin(\beta_n \ell_4) + b_4 \cos(\beta_n \ell_4), \\ -\beta_n b_3 \sin(\beta_n) + \beta_n a_3 \cos(\beta_n) - \beta_n b_4 \sin(\beta_n \ell_4) + \beta_n a_4 \cos(\beta_n \ell_4) = \\ -(i + c_n)\beta_n (a_3 \sin(\beta_n) + b_3 \cos(\beta_n)). \end{cases}$$

$$(5.42)$$

Our goal is to prove that $\beta_n a_3$ converge to infinity. A straightforward calculation leads to

$$(FA + GB)\beta_n a_3 = BH + FC, \qquad (5.43)$$

where

$$A = \sin \beta_n + ((i + c_n) \sin \beta_n + \cos \beta_n) \sin \beta_n \ell_4 + \sin \beta_n \cos(\beta_n \ell_4),$$

$$B = -\cos \beta_n + \left(\frac{\cos^2 \beta_n}{\sin \beta_n} + 3(i + c_n) \cos \beta_n - (2 - c_n^2 - 2i c_n) \sin \beta_n \right) \sin(\beta_n \ell_4)$$

$$+ (2 \cos \beta_n - (i + c_n) \sin \beta_n) \cos(\beta_n \ell_4),$$

$$C = \left(\frac{1}{2}(1 + \frac{i + c_n}{\beta_n}) \sin \beta_n - \frac{i + c_n}{2} \cos \beta_n \right) \sin(\beta_n \ell_4)$$

$$+ \frac{1}{2}(\frac{\sin \beta_n}{\beta_n} - \cos \beta_n) \cos(\beta_n \ell_4),$$

and

$$F = -\sin \beta_n + (i + c_n) \cos \beta_n - (2 \cos \beta_n + (i + c_n) \sin \beta_n) \sin(\beta_n \ell_4)$$

$$+ \left(\frac{\cos^2 \beta_n}{\sin \beta_n} + 3(i + c_n) \cos \beta_n - (2 - c_n^2 - 2i c_n) \sin \beta_n \right) \cos(\beta_n \ell_4),$$

$$G = \cos \beta_n + (i + c_n) \sin \beta_n + \sin \beta_n \sin(\beta_n \ell_4)$$

$$- ((i + c_n) \sin \beta_n + \cos \beta_n) \cos(\beta_n \ell_4),$$

$$H = \frac{1}{2} \left(\frac{\sin \beta_n}{\beta_n} - \cos \beta_n \right) \sin(\beta_n \ell_4) - \left(\frac{1}{2}(1 + \frac{i + c_n}{\beta_n}) \sin \beta_n - \frac{i + c_n}{2} \cos \beta_n \right)$$

$$\times \cos(\beta_n \ell_4).$$

Now, by using the Asymptotic Dirichlet theorem [86], there exists $(P_n, Q_n) \in \mathbb{N}^2$ such that $\frac{P_n}{Q_n}$ converge to ℓ_4, P_n and Q_n tend to infinity as n goes to infinity and for every n in $\mathbb{N}$,

$$|Q_n \ell_4 - P_n| < \frac{1}{Q_n}.$$

Take $\beta_n = 2\pi Q_n + \frac{2\pi}{Q_n^{1/4}}$; then, there exists a positive integer n_0 such that for every integer $n \geq n_0$,

$$0 < \lambda_n := -\frac{2\pi}{Q_n} + \frac{2\pi \ell_4}{Q_n^{1/4}} < \beta_n \ell_4 - 2\pi P_n < \mu_n := \frac{2\pi}{Q_n} + \frac{2\pi \ell_4}{Q_n^{1/4}} < \frac{\pi}{2}$$

and

$$\sin(\lambda_n) < \sin(\beta_n \ell_4) < \sin(\mu_n), \quad \cos(\mu_n) < \cos(\beta_n \ell_4) < \cos(\lambda_n).$$

Moreover, $\sin(\beta_n \ell_4)$ and $\cos(\beta_n \ell_4)$ satisfy the following asymptotic approximations:

$$\sin(\beta_n \ell_4) = \frac{2\pi \ell_4}{\mathcal{Q}_n^{1/4}} + o\left(\frac{1}{\mathcal{Q}_n^{1/4}}\right), \quad \cos(\beta_n \ell_4) = 1 + o\left(\frac{1}{\mathcal{Q}_n^{1/4}}\right).$$

We also have

$$\sin(\beta_n) = \sin\left(\frac{2\pi}{\mathcal{Q}_n^{1/4}}\right) = \frac{2\pi}{\mathcal{Q}_n^{1/4}} + o\left(\frac{1}{\mathcal{Q}_n^{1/4}}\right), \quad \cos(\beta_n) = 1 + o\left(\frac{1}{\mathcal{Q}_n^{1/4}}\right), \quad \text{and}$$

$$\cot(\beta_n) = \frac{\mathcal{Q}_n^{1/4}}{2\pi}\left(1 + o\left(\frac{1}{\mathcal{Q}_n^{1/4}}\right)\right).$$

It follows that

$$A = \frac{2\pi(2+\ell_4)}{\mathcal{Q}_n^{1/4}} + o\left(\frac{1}{\mathcal{Q}_n^{1/4}}\right), \quad B = 1 + \ell_4 + o(1), \quad C = -\frac{1}{2} + o(1),$$

$$F = \frac{\mathcal{Q}_n^{1/4}}{2\pi}\left(1 + 4i\frac{2\pi}{\mathcal{Q}_n^{1/4}} + o\left(\frac{1}{\mathcal{Q}_n^{1/4}}\right)\right), \quad G = o\left(\frac{1}{\mathcal{Q}_n^{1/4}}\right), \quad H = -\frac{i}{2} + o(1).$$

Returning back to (5.43), we could write

$$(2 + \ell_4 + o(1))\beta_n a_3 = -\frac{i}{2}(1+\ell_4) + o(1) - \frac{\mathcal{Q}_n^{1/4}}{4\pi}\left(1 + 4i\frac{2\pi}{\mathcal{Q}_n^{1/4}} + o\left(\frac{1}{\mathcal{Q}_n^{1/4}}\right)\right).$$

Hence,

$$\beta_n a_3 \sim -\frac{1}{4\pi(2+\ell_4)}\mathcal{Q}_n^{1/4},$$

which implies that $\|y_{3,n}\|$ converges to infinity as n goes to infinity and that consequently $zq_{\underline{n}}$ is not bounded. $\qquad\square$

Remark 5.7 A small change in the proof leads to the conclusion that a polynomial stability cannot be better than $\frac{1}{t^2}$ in the case of this special circuit (by using a frequency domain characterization of polynomial stability of a C_0-semigroup of contraction due to Borichev and Tomilov and given in Theorem 1.26). Precisely, we prove that the system is not $\frac{1}{t^\alpha}$-polynomially stable for every α in $(2, \infty)$.

3.2   A Star with Two Fixed Endpoints (Fig. 5.3)

In this example, we have taken $\ell_1 = \ell_2 = 1$, and the two exterior ends of e_1 and e_2 are supposed to be fixed. More precisely, we consider the following system:

$$
\begin{cases}
y_{j,tt} - y_{j,xx} = 0 \text{ in } (0,1) \times (0,\infty), \quad j \in \{1,2\}, \\
y_{3,tt} - y_{3,xx} = 0 \text{ in } (0,\ell_3) \times (0,\infty), \\[6pt]
\displaystyle\sum_{j=1}^{3} y_{j,x}(0,t) = s'(t), \quad s''(t) + s(t) = -y_t(0,t), \\
y_j(0,t) = y_l(0,t), \quad j,l \in \{1,2,3\}, \\
y_2(1,t) = y_3(\ell_3,t) = 0, \quad y_{1,x}(1,t) = -y_t(1,t), \\[6pt]
s(0) = s_0, \quad s'(0) = s_1, \\
y_j(x,0) = y_j^0(x), \quad y_{j,t}(x,0) = y_j^1(x), \quad x \in (0,1), \quad j \in \{1,2\}, \\
y_3(x,0) = y_3^0(x), \quad y_{3,t}(x,0) = y_3^1(x), \quad x \in (0,\ell_3).
\end{cases}
\tag{5.44}
$$

We can rewrite the system (5.44) in the Hilbert space H as

$$
z'(t) = \mathcal{A}z(t),
$$

where

$$
H = V \times \left(\left(L^2(0,1) \right)^2 \times L^2(0,\ell_3) \right) \times \mathbb{C}^2
$$

with

$$
V = \Big\{ \underline{\Phi} \in \left(H^1(0,1) \right)^2 \times H^1(0,\ell_3), \ \Phi_2(1) = \Phi_3(\ell_3) = 0,
$$

$$
\Phi_j(0) = \Phi_l(0), \ j,l \in \{1,2,3\} \Big\}
$$

and

$$
\mathcal{A}\left(\underline{y}, \underline{v}, p, q \right) = \left(\underline{v}, \partial_x^2 \underline{y}, q, -p - \underline{v}(0) \right), \forall \left(\underline{y}, \underline{v}, p, q \right) \in \mathcal{D}(\mathcal{A}),
$$

with

$$
\mathcal{D}(\mathcal{A}) = \Big\{ \left(\underline{y}, \underline{v}, p, q \right) \in \left[\left(\left(H^2(0,1) \right)^2 \times H^2(0,\ell_3) \right) \cap V \right] \times V \times \mathbb{C}^2;
$$

$$
\sum_{j=1}^{3} \partial_x y_j(0) = q, \ \text{and } \partial_x y_1(1) = -v_1(1) \Big\}.
$$

The operator $\mathcal{A}$ generates a C_0-semigroup of contraction $(S(t))_{t\geq 0}$, and we have the following result.

Theorem 5.8 *The semigroup $(S(t))_{t\geq 0}$ is asymptotically stable if and only if ℓ_3 is irrational and not in $\pi\mathbb{Z}$.*
Even if ℓ_3 is irrational and not in $\pi\mathbb{Z}$, the semigroup $(S(t))_{t\geq 0}$ is not exponentially stable.

Proof As in the case of a circuit, we consider the sequence f_n of vectors of H defined by $f_n = (0, \underline{g}_n, 0, 0)$, where $\underline{g}_n = (0, -\sin\beta_n x, 0)$ and β_n is a sequence of real numbers satisfying $\beta_n \longmapsto +\infty$ and that will be defined later, such that the sequence $z_n = (\underline{y}_n, \underline{v}_n, p_n, q_n)$ of elements of $\mathcal{D}(\mathcal{A})$ satisfying $(i\beta_n - \mathcal{A})z_n = f_n$ is not bounded.

The sequences $\underline{y}_n$ and q_n should satisfy

$$
\begin{cases}
\beta_n^2 y_{j,n} + \partial_x^2 y_{j,n} = 0, & \text{for } j = 1, 3, \\
\beta_n^2 y_{2,n} + \partial_x^2 y_{2,n} = \sin\beta_n x, \\
q_n = -\dfrac{\beta_n^2}{\beta_n^2 - 1}\underline{y}_n(0) = -\beta_n c_n \underline{y}_n(0),
\end{cases}
$$

with $c_n = \dfrac{\beta_n}{\beta_n^2 - 1}$. Then, for $j = 1, 3$, there exist two complex numbers a_j and b_j (depending of n) such that

$$
\begin{cases}
y_{j,n} = a_j \sin(\beta_n x) + b_j \cos(\beta_n x), \\
\partial_x y_{j,n} = -\beta_n b_j \sin(\beta_n x) + \beta_n a_j \cos(\beta_n x),
\end{cases}
$$

and there exist two complex numbers a_2 and b_2 (depending of n) such that

$$
\begin{cases}
y_n^2 = a_2 \sin(\beta_n x) + (-\frac{x}{2\beta_n} + b_2)\cos(\beta_n x), \\
\partial_x y_{2,n} = (\frac{x}{2} - \beta_n b_2)\sin(\beta_n x) + (-\frac{1}{2\beta_n} + \beta_n a_2)\cos(\beta_n x).
\end{cases}
$$

The boundary and transmission conditions are expressed as follows:

$$
\begin{cases}
a_2 \sin(\beta_n) + (-\frac{1}{2\beta_n} + b_2)\cos(\beta_n) = 0, \\
a_3 \sin(\beta_n \ell_3) + b_1 \cos(\beta_n \ell_3) = 0, \\
b_1 = b_2 = b_3, \\
-\frac{1}{2\beta_n} + \beta_n a_1 + \beta_n a_2 + \beta_n a_3 = -\beta_n c_n b_1, \\
-\beta_n b_1 \sin(\beta_n) + \beta_n a_1 \cos(\beta_n) = -\beta_n a_1 \sin(\beta_n) - i\beta_n b_1 \cos(\beta_n).
\end{cases}
\tag{5.45}
$$

From the first, the second, and the sixth equations in (5.45), we deduce

$$
a_2 = \left(\frac{1}{2\beta_n} - b_2\right)\cot(\beta_n), \quad a_3 = -b_3 \cot(\beta_n \ell_3), \quad \text{and} \quad a_1 = -ib_1.
\tag{5.46}
$$

Using (5.46) in the fifth equation of (5.45), taking into account that $b_1 = b_2 = b_3$, to get

$$\beta_n b_1 \left[-c_n + i + \cot(\beta_n) + \cot(\beta_n \ell_3) \right] = -\frac{1}{2\beta_n} + \frac{1}{2} \cot(\beta_n). \qquad (5.47)$$

As in the previous case, there exists $(P_n, Q_n) \in \mathbb{N}^2$ such that $\frac{P_n}{Q_n}$ converge to ℓ_3, P_n and Q_n tend to infinity as n goes to infinity and for every n in $\mathbb{N}$,

$$|Q_n \ell_4 - P_n| < \frac{1}{Q_n}.$$

We take again $\beta_n = 2\pi Q_n + \frac{2\pi}{Q_n^{1/4}}$, and then we get (as in the first case)

$$\cot(\beta_n \ell_3) = \frac{Q_n^{1/4}}{2\pi \ell_3} \left(1 + o\left(\frac{1}{Q_n^{1/4}} \right) \right), \quad \text{and} \quad \cot(\beta_n) = \frac{Q_n^{1/4}}{2\pi} \left(1 + o\left(\frac{1}{Q_n^{1/4}} \right) \right).$$

It follows from (5.47) that

$$\beta_n b_1 (1 + \ell_3) \frac{Q_n^{1/4}}{2\pi \ell_3} \sim \frac{Q_n^{1/4}}{4\pi \ell_3}$$

as n goes to infinity. Hence, using the second equality in (5.46), we obtain

$$\beta_n a_3 \sim -\frac{1}{4\pi(1 + \ell_3)} Q_n^{1/4}$$

as n goes to infinity, which implies that $\| y_{3,n} \|$ converges to infinity as n goes to infinity and that consequently z_n is not bounded. $\square$

4   A Chain with Non-equal Mass Points

In this section, we consider a particular network which is a chain of N edges ($N \geq 2$) and $p = N+1$ vertices such that the $(N-1)$ interior vertices a_j are point masses with mass m_j. But the masses m_j are not necessarily equal (Fig. 5.4).

Precisely, we consider the following system:

$$\begin{cases} y_{j,tt} - y_{j,xx} = 0 \text{ in } (0, \ell_j) \times (0, \infty), \quad j \in \{1, \ldots, N\}, \\ \\ y_{j,x}(0, t) - y_{j-1,x}(\ell_{j-1}, t) = s'_j(t), \quad j \in \{2, \ldots, N\}, \\ m_j s''_j(t) + s_j(t) = -y_t(0, t), \quad j \in \{2, \ldots, N\}, \\ y_{j-1}(\ell_{j-1}, t) = y_j(0, t), \quad j \in \{2, \ldots, N\}, \\ y_{1,x}(0, t) = y_t(0, t), \\ y_N(\ell_N, t) = 0, \\ \\ s_j(0) = s_j^0, \quad s'_j(0) = s_j^1, \quad j \in \{2, \ldots, N\}, \\ y_j(x, 0) = y_j^0(x), \quad y_{j,t}(x, 0) = y_j^1(x), \quad x \in (0, \ell_j), \quad j \in \{1, \ldots, N\}. \end{cases} \quad (5.48)$$

Note that the feedback is applied at the vertex a_1. We give a necessary and sufficient condition for the exponential stability of system (5.48). The general case of a tree with distinct masses at inner nodes is complicated for the moment, because the calculations are based on some recurrence relations, something we could not do for a general tree.

To start, we quickly redefine the associated state space H and the operator $\mathcal{A}$ as follows:

$$H = V \times \prod_{j=1}^{N} L^2(0, \ell_j) \times \left(\prod_{j=1}^{N-1} \mathbb{C} \right)^2$$

with

$$V = \left\{ \Phi \in \prod_{j=1}^{N} H^1(0, \ell_j), \ \Phi_N(\ell_N) = 0, \ \Phi_{j-1}(\ell_{j-1}) = \Phi_j(0), \ j = 2, \ldots, N \right\}$$

and

$$\mathcal{A}\left(\underline{y}, \underline{v}, \underline{p}, \underline{q}\right) = \left(\underline{y}, \partial_x^2 \underline{y}, \underline{q}, -m^{-1}(\underline{p} + \underline{v}_\mathcal{M})\right), \ \forall \left(\underline{y}, \underline{v}, \underline{p}, \underline{q}\right) \in \mathcal{D}(\mathcal{A})$$

with $-m^{-1}(\underline{p} + \underline{v}_\mathcal{M}) = (-\frac{1}{m_j} p_j - \frac{1}{m_j} \underline{v}(\ell_j))_{j \in \{2, \ldots, N-1\}}$ and

$$\mathcal{D}(\mathcal{A}) = \left\{ (\underline{y}, \underline{v}, \underline{p}, \underline{q}) \in \left[\prod_{j \in J} H^2(0, \ell_j) \cap V \right] \times V \times \left(\prod_{k \in I_{int}} \mathbb{C} \right)^2 \right.$$

$$\left. \text{satisfying (5.49)} \right\},$$

where

$$\begin{cases} \partial_x y_j(0) - \partial_x y(\ell_{j-1}) = q_j, & j \in \{2, \dots, N\}, \\ \partial_{xy_1}(0) = v_1(0). \end{cases} \tag{5.49}$$

Then, the operator $\mathcal{A}$ generates a $\mathcal{C}_0$-semigroup of contraction $(S(t))_{t \geq 0}$. Moreover, $\sigma(\mathcal{A}) = \sigma_p(\mathcal{A})$.

For every mass point m, we denote by $i_1(m), \dots, i_{k_m}(m)$ the indices of the interior nodes with masses equal to m and ordered as follows: $i_1(m) < i_2(m) < \dots < i_{k_m}(m)$.

For $r = r(m) \in \{1, \dots, k_m\}$, we define the scalars

$$\Pi_{m,r(m),s} = \sum_{i_r = j_0 < j_1 < \dots < j_s < i_{r+1}} \left(\prod_{i=0}^{s-1} c_{j_{i+1}} \sin(\beta \ell_{j_i} + \dots + \beta \ell_{j_{i+1}-1}) \right)$$
$$\times \sin(\beta \ell_{j_s} + \dots + \beta \ell_{i_{r+1}-1}),$$

where $\beta = \pm \frac{1}{m}$ and $c_{j_i} = \frac{1}{\beta(m_{j_i} - m)}$ for $i = 1, \dots, s$ and

$$\Delta_{r(m)} := (-1)^{i_{r+1} - i_r} \sin(\beta \ell_{i_r} + \dots + \beta \ell_{i_{r+1}-1})$$
$$+ \sum_{s=0}^{i_{r+1} - i_r - 1} (-1)^{i_{r+1} - i_r + s} \Pi_{m,r(m),s},$$

with $\Delta_{r(m)} = \sin(\beta \ell_r)$ if $k_m = 1$.

Then, we have the following result of asymptotic behavior of the system (5.48):

Lemma 5.9 *The system defined by (5.48) is exponentially stable if and only if for every mass point m and for every $r(m)$, $\Delta_{r(m)} \neq 0$.*

Proof The first question is whether $i\mathbb{R}$ belongs to $\rho(A)$. Thus, we will solve the equation

$$\mathcal{A}z = i\beta z \tag{5.50}$$

of unknown $z = \left(y, \underline{v}, \underline{p}, \underline{q} \right) \in \mathcal{D}(\mathcal{A}) - \{0\}$ and $\beta \in \mathbb{R}$.

If $\beta = 0$, then $z = 0$. Thus, suppose that $\beta \neq 0$. We have

$$v_j = i\beta y_j, \quad \beta^2 y_j + \partial_x^2 y_j dx^2 = 0, \quad j \in \{1, \dots, N\}, \tag{5.51}$$

$$q_j = i\beta p_j, \quad (m_j \beta^2 - 1)p_j = \underline{v}(a_j), \quad j \in \{2, \dots, N\}. \tag{5.52}$$

The function y_j, $j \in \{1, \ldots, N\}$, is then of the form

$$y_j = \alpha_j \cos(\beta x) + \gamma_j \sin(\beta x).$$

If $m_j \beta^2 \neq 1$ for every j in $\{2, \ldots, N\}$, then we prove by iteration, starting with $j = 1$, that $y_j = 0$ for every j in $\{2, \ldots, N\}$ and consequently $z = 0$.

Now, we suppose the existence of a mass point m with $m\beta^2 = 1$. Let $i_1(m) < \ldots < i_{k_m}(m)$ be the indices of inner nodes with masses equal to m. Then, as in the first case, $y^j = 0$ for every $j < i_1$. Let $r = r(m) \in \{1, \ldots, k_m\}$, and we have the following system:

$$(5.53) \quad \begin{cases} y_r(0) = 0, \\ \text{for } j = i_r \text{ to } j = i_{r+1} - 2, \ y_j(\ell_j) = y_{j+1}(0), \\ \text{and } -\partial_x y_j(\ell_j) + \partial_x y_{j+1}(0) = q_{j+1}, \\ y_{i_{r+1}-1}(\ell_{i_{r+1}-1}) = 0. \end{cases}$$

It is obvious that the system (5.50) has trivial solution if and only if for every m and every $r(m)$ the system (5.53) has a trivial solution.

By changes of indices, we can suppose that $i_r = 2$, $i_{r+1} = N + 1$. The system (5.53) can be rewritten as

$$\begin{cases} \alpha_2 = 0, \\ \text{for } j = 2 \text{ to } j = N - 1, \ \alpha_j \cos(\beta \ell_j) + \gamma_j \sin(\beta \ell_j) = \alpha_{j+1}, \\ \text{and } \alpha_j \sin(\beta \ell_j) - \gamma_j \cos(\beta \ell_j) + \frac{1}{\beta} \frac{1}{m_j - m} \alpha_{j+1} + \gamma_{j+1} = 0, \\ \alpha_N \cos(\beta \ell_N) + \gamma \sin(\beta \ell_N) = 0. \end{cases}$$

The matrix of such system is

$$S_N = \begin{pmatrix}
1 & 0 & 0 & & & & & & \\
\cos x_2 & \sin x_2 & -1 & 0 & & & & & \\
\sin x_2 & -\cos x_2 & c_3 & 1 & 0 & & & & \\
0 & 0 & \cos x_3 & \sin x_3 & -1 & 0 & & (0) & \\
0 & 0 & \sin x_3 & -\cos x_3 & c_4 & 1 & 0 & & \\
& & \cdots\cdots & \cdots & \cdots & & & \\
& & \cdots\cdots & \cdots & \cdots & & 0 & \\
& (0) & & \cdots & \cos x_{N-1} & \sin x_{N-1} & -1 & 0 \\
& & & \cdots & \sin x_{N-1} & -\cos x_{N-1} & c_N & 1 \\
& & & \cdots & 0 & 0 & \cos x_N & \sin x_N
\end{pmatrix},$$

where $x_j = \beta \ell_j$ and $c_j = \frac{1}{\beta(m_j - m)}$. We want to calculate the determinant Δ_N of S_N. For this, let M_N be the determinant of the matrix obtained from S_N by replacing $\cos x_N$ and $\sin x_N$ in the last line by $\sin x_N$ and $-\cos x_N$, respectively.

One can verify easily that

$$\Delta_2 = \sin x_2, \quad \Delta_3 = -\sin(x_2 + x_3) + c_3 \sin x_2 \sin x_3,$$

$$\Delta_4 = \sin(x_2 + x_3 + x_4) - c_3 \sin x_2 \sin(x_3 + x_4) - c_4 \sin(x_2 + x_3) \sin x_4$$

$$+ c_3 c_4 \sin x_2 \sin x_3 \sin x_4,$$

$$M_2 = -\cos x_2, \quad M_3 = \cos(x_2 + x_3) - c_3 \sin x_2 \cos x_3,$$

$$M_4 = -\cos(x_2 + x_3 + x_4) + c_3 \sin x_2 \cos(x_3 + x_4) + c_4 \sin(x_2 + x_3) \cos x_4$$

$$-c_3 c_4 \sin x_2 \sin x_3 \cos x_4.$$

We will prove by induction that for every integer $N \geq 2$,

$$\Delta_N = (-1)^N \sin(x_2 + \ldots + x_N) + (-1)^{N-1}$$

$$\times \sum_{j=2}^{N-1} c_{j+1} \sin(x_2 + \ldots + x_{j-1}) \sin(x_j + \ldots + x_N)$$

$$+ \sum_{s=2}^{N-2} (-1)^{N-1-s} \sum_{2=j_0<j_1<\cdots<j_s\leq N} \left(\prod_{i=0}^{s-1} c_{j_{i+1}} \sin(x_{j_i} + \ldots + x_{j_{i+1}-1}) \right)$$

$$\times \sin(\beta\ell_{j_s} + \cdots + \beta\ell_N) \tag{5.54}$$

and

$$M_N = (-1)^{N+1} \cos(x_2 + \ldots + x_N) + (-1)^N$$

$$\times \sum_{j=2}^{N-1} c_{j+1} \sin(x_2 + \ldots + x_{j-1}) \cos(x_j + \ldots + x_N)$$

$$+ \sum_{s=2}^{N-2} (-1)^{N-s} \sum_{2=j_0<j_1<\cdots<j_s\leq N} \left(\prod_{i=0}^{s-1} c_{j+1} \sin(x_{j_i} + \ldots + x_{j_{i+1}-1}) \right)$$

$$\times \cos(x_{j_s} + \ldots + x_N). \tag{5.55}$$

Such rules are true for $N = 2$ and $N = 3$. Let $N \in \mathbb{N}$ with $N \geq 3$, and suppose that (5.54) and (5.55) are true. Some calculations lead to

$$\Delta_N = (-\cos x_N + c_N \sin x_N)\Delta_{N-1} + (\sin x_N)M_{N-1},$$

$$M_N = (-\sin x_N - c_N \cos x_N)\Delta_{N-1} - (\cos x_N)M_{N-1}.$$

We can now verify the rule (5.54) to order N

$$\Delta_N = \left((-1)^N \sin(x_2 + \ldots + x_N) + (-1)^{N-1} \sum_{j=2}^{N-1} c_{j+1} \sin(x_2 + \ldots + x_{j-1}) \sin(x_j + \ldots + x_N) \right.$$

$$+ \sum_{s=2}^{N-3} (-1)^{N-1-s} \sum_{2=j_0\leq j_1<\cdots<j_s\leq N-1} \left(\prod_{i=0}^{s-1} c_{j_{i+1}} \sin(x_{j_i} + \ldots + x_{j_{i+1}-1}) \right) \sin(x_{j_s} + \ldots + x_N) \right)$$

$$+ \sum_{s=2}^{N-3} (-1)^{N-2-s} \sum_{2=j_0\leq j_1<\cdots<j_s\leq N-1} c_N \left(\prod_{i=0}^{s-1} c_{j_{i+1}} \sin(x_{j_i} + \ldots + x_{j_{i+1}-1}) \right) \sin(x_{j_s} + \ldots + x_{N-1}) \sin x_N \right)$$

$$
= \left((-1)^N \sin(x_2 + \ldots + x_N) + (-1)^{N-1} \sum_{j=2}^{N-1} c_{j+1} \sin(x_2 + \ldots + x_{j-1}) \sin(x_j + \ldots + x_N) \right.
$$

$$
\left. + \sum_{s=2}^{N-2} (-1)^{N-1-s} \sum_{2=j_0 \leq j_1 < \cdots < j_s \leq N} \left(\prod_{i=0}^{s-1} c_{j_{i+1}} \sin(x_{j_i} + \ldots + x_{j_{i+1}-1}) \right) \sin(x_{j_s} + \ldots + x_N) \right).
$$

A similar calculus, using (5.55), proves that (5.55) is verified in order N.

We can now state the following result.

The associated semigroup $S(t)$ is asymptotically stable if and only if $\Delta_{r(m)}$ is different from zero for every mass point m and every $r(m)$. To conclude that $(S(t))_{t \geq 0}$ is exponentially stable, it suffices to prove that (1.10) is satisfied by $(S(t))_{t \geq 0}$, exactly as in the proof of Theorem 5.4   □

Chapter 6
Stability of a Graph of Strings with Local Kelvin–Voigt Damping

Viscoelastic materials, as their name suggests, combine two different properties: viscosity and elasticity. They are used for isolating vibration, dampening noise, and absorbing shock. They are intended to dissipate mechanical energy from vibrations or noises; to limit their propagation in structures, they have a decisive impact on the fatigue of these structures and on our comfort.

Viscoelastic materials have applications in all fields of engineering and mechanical systems, from the automotive to civil engineering, from space to home appliances (engine and machine mounts and supports, transmission seals and belts, glazing edges and fixing of subsystems, damping of metal plates and shells, parts of seats and interior of cabs, tire and wheels, tuned damping systems...) [27, 82].

Since the 1980s, the development of modern technologies has required the use of innovative materials with high mechanical properties, suitable for their use, and having low densities. A composite material meets most of these requirements; it is a kind of mixture of different materials whose properties are superior to each of its components taken separately. These materials were first developed and used in the 1940s in the aeronautical field (essentially for military airplanes and helicopters) and are today in automobile construction, shipbuilding, and in buildings. But these materials are excellent transmitters of mechanical and acoustic vibrations, which can affect the integrity of the entire system. Also, thanks to these composite materials it is possible to reduce the number of parts of a structure, there would then be less frictions at connections between elements. It is therefore imperative to associate with these materials effective damping techniques. One solution is to add full or partial layers of viscoelastic materials, glued on (or incarnated between) the parts. A viscoelastic product can be integrated into the composite material, [29, 44].

In this context we have chosen to study a network of elastic and viscoelastic materials. More precisely, we investigate the asymptotic stability of a graph of elastic strings with local Kelvin–Voigt damping.

K. Ammari, F. Shel, *Stability of Elastic Multi-Link Structures*, SpringerBriefs in Mathematics, https://doi.org/10.1007/978-3-030-86351-7_6

Models of the transient behavior of some or all of the state variables describing the motion of flexible structures have been of great interest in recent years; for more details about physical motivation for the models, see also [18, 49] and the references therein. Mathematical analysis of transmission partial differential equations is detailed in [49]. For the feedback stabilization problem of the wave or Schrödinger equations in networks, we refer the readers to references [4, 9–11, 49].

A wave equation on a (single) string of length ℓ with (local) Kelvin–Voigt damping is modeled by the following equation:

$$\frac{\partial^2 u}{\partial t^2} - \frac{\partial}{\partial x}\left(\frac{\partial u}{\partial x} + \beta(x)\frac{\partial^2 u}{\partial x \partial t}\right) = 0 \quad \text{in} \quad (0, \ell) \times (0, \infty), \tag{6.1}$$

where $\beta(x), x \in [0, \ell]$ is a nonnegative function.

As boundary conditions, we often associate the Dirichlet conditions:

$$u(0, t) = u(\ell, t) = 0.$$

From a mathematical point of view, the Kelvin–Voigt damping model (6.1) has been studied by several authors. Let us recall some results in the literature,

- Huang proved in 1988 [43] that when the damping is global (i.e., distributed over the entire domain), the corresponding semigroup is not only exponentially stable but also analytic. Thus, the Kelvin–Voigt damping is much stronger than the viscous damping (i.e., the damping term is replaced by $-\beta(x)\frac{\partial u}{\partial t}$), where the corresponding semigroup is only exponentially stable and not analytic (see, e.g., [50, 73]).

    Such a comparison is not valid anymore if the damping is localized:
- Chen et al. [73] proved in 1991 that in the case of localized viscous damping, the associated semigroup is exponentially stable no matter the size or the location of the subinterval where the damping is effective, and even if the damping coefficient function has a jump discontinuity at the interface.

However, the local Kelvin–Voigt damping does not follow the same analogue.

- It was first proved in 1998 by S. Chen et al. [53] that, when the viscoelastic damping is locally distributed (precisely, they took $\beta(x) = \beta_0 \chi_{(\alpha, w)}$, with $\beta_0 > 0$), the associated semigroup is not exponentially stable.
- In 2002, K. Liu and Z. Liu [54] proved that if $\beta \in C^2[0, \ell]$, and $\int_0^\ell \beta(x)dx > 0$, then the system is exponentially stable: The asymptotic behavior depends on the regularity of the damping coefficient.

The works cited below consider the domain $[-1, 1]$ instead of $[0, \ell]$ and suppose that $\beta(x) = 0$ on $[-1, 0]$ and $\beta(x) = b(x)$ on $(0, 1]$.

- In 2004, Renardy [83] supposed that $\beta(x) = 0$ on $[-1, 0]$ and $\beta(x) > 0$ on $(0, 1]$ and he assumed that

$$\lim_{x \to 0^+} \frac{\beta'(x)}{x^\alpha} = k > 0 \quad \text{for some} \ \alpha > 0, \tag{6.2}$$

then the eigenvalues of the system (6.1) are such that the decay rate tends to infinity with frequency.

- Z. liu and B. Rao [55], 2005, and M. Alves et al. [87], 2014, proved that if $b(x) \geq c > 0$ on $(0, 1)$ and $b \in \mathcal{C}(0, 1)$, the associated semigroup is polynomially stable of order 2.

- In 2010, Q. Zhang [104] improved the result in [55]: The author took $\beta \in \mathcal{C}^1[-1, 1]$, $b(0) = b'(0) = 0$ and supposed the existence of a positive constant c such that $\int_0^x \frac{|b'(s)|^2}{b(s)} ds \leq c|b'(x)|$ for all $x \in [0, 1]$, (for example, $b(x) = x^\alpha$, $\alpha > 1$).

- In 2016 Z. Liu and Q. Zhang [60] took over the condition (6.2) of Renardy. Precisely they took $\beta \in L^\infty(-1, 1)$, $b(x) > 0$ on $(0, 1]$ and $b(0) = 0$; $b', b'' \in L^\infty(0, 1)$, and supposed that $\lim_{x \to 0^+} \frac{\beta(x)}{x^\alpha} = k > 0$. Then the system (6.1) is exponentially stable for $\alpha = 1$ and polynomially, non-exponentially stable for $0 \leq \alpha < 1$.

- It is proved [56] in 2017 that if $\beta \in \mathcal{C}^1[-1, 1]$ and satisfies conditions in the last point, then the system (6.1) remains exponentially stable for $\alpha > 1$.

In this work we study a more general case, and it is about a network of strings with local Kelvin–Voigt damping.

Recall that $\mathcal{G}$ denotes a planar metric connected graph embedded in $\mathbb{R}^m$, $m \in \mathbb{N} \setminus \{0\}$, with N edges $e_1, \ldots, e_N$, $N \geq 1$ and p vertices $a_1, \ldots, a_p$, $p \geq 2$. The incidence matrix $D = (d_{kj})_{p \times N}$, is defined by,

$$d_{kj} = \begin{cases} 1 & \text{if} \ \pi_j(\ell_j) = a_k, \\ -1 & \text{if} \ \pi_j(0) = a_k, \\ 0 & \text{otherwise.} \end{cases}$$

Suppose that the equilibrium position of our network of elastic strings coincides with the graph $\mathcal{G}$. Then, we consider the following initial and boundary value problem:

$$\frac{\partial^2 u_j}{\partial t^2}(x, t) - \frac{\partial}{\partial x}\left(\frac{\partial u_j}{\partial x} + \beta_j(x)\frac{\partial^2 u_j}{\partial x \partial t}\right)(x, t) = 0, \quad 0 < x < \ell_j, \ t > 0, \ j \in J,$$

$$\tag{6.3}$$

$$u_{j_k}(a_k, t) = 0, \ a_k \in \mathcal{V}_{ext}, \ t > 0, \tag{6.4}$$

$$u_j(a_k, t) = u_l(a_k, t), \quad t > 0, \ j, l \in J_k, \ a_k \in \mathcal{V}_{int}, \tag{6.5}$$

Fig. 6.1 A graph

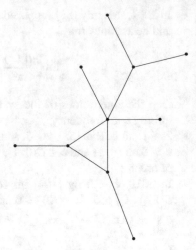

$$\sum_{j \in J_k} d_{kj} \left(\frac{\partial u_j}{\partial x}(a_k, t) + \beta_j(a_k) \frac{\partial^2 u_j}{\partial x \partial t}(a_k, t) \right) = 0, \quad t > 0, \; a_k \in \mathcal{V}_{int}, \tag{6.6}$$

$$u_j(x, 0) = u_j^0(x), \quad \frac{\partial u_j}{\partial t}(x, 0) = u_j^1(x), \quad 0 < x < \ell_j, \; j \in J, \tag{6.7}$$

where $u_j : [0, \ell_j] \times (0, +\infty) \to \mathbb{R}$, $j \in J$, be the transverse displacement in e_j, $\beta_j \in L^\infty(0, \ell_j)$ and, either β_j is zero, that is, e_j is a purely elastic edge, or there exists a subinterval Ω_j of $(0, \ell_j)$, not reduced to a singleton, such that $\beta_j(x) > 0$, a.e. on Ω_j. Such edge will be called a K-V edge (Fig. 6.1).

We assume that $\mathcal{G}$ contains at least one K-V edge and contains at least one external node (i.e., $I_{ext} \neq \varnothing$). Furthermore, we suppose that every maximal subgraph of purely elastic edges is a tree, whose leaves are attached to K-V edges.

Our aim is to prove, under some assumptions on damping coefficients β_j, $j \in J$, exponential and polynomial stability results for the system (6.3)–(6.7).

We define the natural energy $E(t)$ of a solution $\underline{u} = (u_j)_{j \in J}$ of (6.3)–(6.7) by

$$E(t) = \frac{1}{2} \sum_{j \in J} \int_0^{\ell_j} \left(\left| \frac{\partial u_j}{\partial t}(x, t) \right|^2 + \left| \frac{\partial u_j}{\partial x}(x, t) \right|^2 \right) dx. \tag{6.8}$$

It is straightforward to check that every sufficiently smooth solution of (6.3)–(6.7) satisfies the following dissipation law:

$$\frac{dE}{dt}(t) = - \sum_{j \in J} \int_0^{\ell_j} \beta_j(x) \left| \frac{\partial^2 u_j}{\partial x \partial t}(x, t) \right|^2 dx \le 0, \tag{6.9}$$

and therefore, the energy is a nonincreasing function of the time variable t.

The main results of this paper then concern the precise asymptotic behavior of the solutions of (6.3)–(6.7). Our technique is a special frequency domain analysis of the corresponding operator.

Recall that, the notation $u_n = o(1)$ and $u_n = O(1)$ for a sequence u_n of complex numbers means $u_n \to 0$ as $n \to \infty$ and u_n is bounded respectively.

This work is organized as follows: In Sect. 1, we give the proper functional setting for system (6.3)–(6.7) and prove that the system is well-posed. In Sect. 2, we analyze the resolvent of the wave operator associated to the dissipative system (6.3)–(6.7) and prove the asymptotic behavior of the corresponding semigroup. For more details in the proofs, see [8].

1 Well-Posedness of the System

In order to study system (6.3)–(6.7) we need a proper functional setting. We define the following space:

$$\mathcal{H} = V \times H,$$

where

$$H = \prod_{j \in J} L^2(0, \ell_j)$$

and

$$V = \left\{ \underline{u} \in \prod_{j \in J} H^1(0, \ell_j) : u_{j_k}(a_k) = 0, \ a_k \in \mathcal{V}_{ext}, \ \text{satisfies (6.10)} \right\}$$

$$u_j(a_k) = u_l(a_k) := \underline{u}(a_k), \ a_k \in \mathcal{V}_{int}, \quad j, l \in J_k, \tag{6.10}$$

and equipped with the inner products

$$< (\underline{u}, \underline{v}, (\tilde{\underline{u}}, \tilde{\underline{v}}) >_{\mathcal{H}} = \sum_{j \in J} \int_0^{\ell_j} \left(v_j(x) \overline{\tilde{v}}_j(x) + \partial_x u_j(x) \, \partial_x \overline{\tilde{u}}(x) \right) dx. \tag{6.11}$$

System (6.3)–(6.7) can be rewritten as the first-order evolution equation

$$\begin{cases} \dfrac{\partial}{\partial t} \begin{pmatrix} \underline{u} \\ \frac{\partial \underline{u}}{\partial t} \end{pmatrix} = \mathcal{A} \begin{pmatrix} \underline{u} \\ \frac{\partial \underline{u}}{\partial t} \end{pmatrix}, \\[2mm] \underline{u}(0) = \underline{u}^0, \ \dfrac{\partial \underline{u}}{\partial t} = \underline{u}^1 \end{cases} \tag{6.12}$$

where the operator $\mathcal{A} : \mathcal{D}(\mathcal{A}) \subset \mathcal{H} \to \mathcal{H}$ is defined by

$$\mathcal{A}\begin{pmatrix} \underline{u} \\ \underline{v} \end{pmatrix} := \begin{pmatrix} \underline{v} \\ \partial_x(\partial_x \underline{u} + \underline{\beta} * \partial_x \underline{v}) \end{pmatrix},$$

with

$$\underline{\beta} := (\beta_j)_{j \in J} \quad \text{and} \quad \underline{\beta} * \partial_x \underline{v} := (\beta_j \partial_x v_j)_{j \in J},$$

and

$$\mathcal{D}(\mathcal{A}) := \left\{ (\underline{u}, \underline{v}) \in \mathcal{H}, \ \underline{v} \in V, \ (\partial_x \underline{u} + \underline{\beta} * \partial_x \underline{v}) \right.$$

$$\left. \in \prod_{j \in J} H^1(0, \ell_j) \ : (\underline{u}, \underline{v}) \text{ satisfies } (6.13) \right\},$$

$$\sum_{j \in J_k} d_{kj} \left(\partial_x u_j(a_k) + \beta_j(a_k) \partial_x v_j(a_k) \right) = 0, \quad t > 0, \ a_k \in \mathcal{V}_{int}. \tag{6.13}$$

Lemma 6.1 *The operator $\mathcal{A}$ is dissipative, $0 \in \rho(\mathcal{A})$: the resolvent set of $\mathcal{A}$.*

Proof For $(\underline{u}, \underline{v}) \in \mathcal{D}(\mathcal{A})$, we have

$$Re(\langle \mathcal{A}(\underline{u}, \underline{v}), (\underline{u}, \underline{v}) \rangle_{\mathcal{H}})$$

$$= Re \sum_{j \in J} \left(\int_0^{\ell_j} \partial_x v_j \overline{\partial_x u_j} dx + \int_0^{\ell_j} \partial_x(\partial_x u_j + \beta_j \partial_x v_j) \overline{v_j} dx \right).$$

Performing integration by parts and using transmission and boundary conditions, a straightforward calculation leads to

$$Re(\langle \mathcal{A}(\underline{u}, \underline{v}), (\underline{u}, \underline{v}) \rangle_{\mathcal{H}}) = - \sum_{j \in J} \int_0^{\ell_j} \beta_j(x) \left| \partial_x v_j(x) \right|^2 dx \leq 0$$

which proves the dissipativeness of the operator $\mathcal{A}$ in $\mathcal{H}$.

Next, using Lax–Milgram's lemma, we prove that $0 \in \rho(\mathcal{A})$. For this, let $(\underline{f}, \underline{g}) \in \mathcal{H}$ and we look for $(\underline{u}, \underline{v}) \in \mathcal{D}(\mathcal{A})$ such that

$$\mathcal{A}(\underline{u}, \underline{v}) = (\underline{f}, \underline{g})$$

which can be written as

$$v_j = f_j, \quad j \in J, \tag{6.14}$$

$$\partial_x(\partial_x u_j + \beta_j \partial_x v_j) = g_j, \quad j \in J. \tag{6.15}$$

$\underline{v}$ is completely determined by (6.14). Let $\underline{w} \in V$; multiplying (6.15) by w_j, then summing over $j \in J$, we obtain, using transmission and boundary conditions,

$$\sum_{j \in J} \int_0^{\ell_j} \left(\partial_x u_j + \beta_j \partial_x v_j\right) \partial_x \overline{w_j} dx = -\sum_{j \in J} \int_0^{\ell_j} g_j \overline{w_j} dx. \tag{6.16}$$

Replacing v_j in the last equality by (6.14), we get

$$\varphi(\underline{u}, \underline{w}) = \psi(\underline{w}), \tag{6.17}$$

where

$$\varphi(\underline{u}, \underline{w}) = \sum_{j \in J} \int_0^{\ell_j} \partial_x u_j \partial_x \overline{w_j}$$

and

$$\psi(\underline{w}) = -\sum_{j \in J} \left(\int_0^{\ell_j} g_j \, \overline{w_j} dx + \int_0^{\ell_j} \beta_j \partial_x f_j \partial_x \overline{w_j} dx \right).$$

The function φ is a continuous sesquilinear form on $V \times V$ and ψ is a continuous anti-linear form on V; here V is equipped with the inner product

$$\left\langle \underline{f}, \underline{g} \right\rangle = \sum_{j \in I} \int_0^{\ell_j} \partial_x f_j \partial_x \overline{g_j}.$$

Since φ is coercive on V, by the Lax–Milgram lemma, Eq. (6.17) has a unique solution $\underline{u} \in V$. Then taking $\underline{w} \in \prod_{j \in J} \mathcal{D}(0, \ell_j)$ in (6.17) and integrating by parts, we deduce that $(\partial_x \underline{u} + \underline{\beta} * \partial_x \underline{v}) \in \prod_{j \in J} H^1(0, \ell_j)$ and $(\underline{u}, \underline{v})$ satisfies (6.14) and (6.15). Moreover $(\underline{u}, \underline{v})$ satisfies (6.13).

Returning back to the Lax–Milgram lemma, $(\underline{u}, \underline{v})$ verifies

$$\left\| (\underline{u}, \underline{v}) \right\|_{\mathcal{H}} \leq \left\| (\underline{f}, \underline{g}) \right\|_{\mathcal{H}}.$$

In conclusion $(\underline{u}, \underline{v}) \in \mathcal{A}$ and $\mathcal{A}^{-1} \in \mathcal{L}(\mathcal{H})$, which assert that $0 \in \rho(\mathcal{A})$. $\qquad \square$

By the Lumer–Phillip's theorem (see [79, 92]), we have the following proposition.

Proposition 6.2 *The operator $\mathcal{A}$ generates a C_0-semigroup of contraction $(S_d(t))_{t\geq 0}$ on the Hilbert space $\mathcal{H}$.*

Hence, for an initial datum $(\underline{u}^0, \underline{u}^1) \in \mathcal{H}$, there exists a unique solution $\left(\underline{u}, \frac{\partial \underline{u}}{\partial t}\right) \in C([0, +\infty), \mathcal{H})$ to problem (6.12). Moreover, if $(\underline{u}^0, \underline{u}^1) \in \mathcal{D}(\mathcal{A})$, then

$$\left(\underline{u}, \frac{\partial \underline{u}}{\partial t}\right) \in C([0, +\infty), \mathcal{D}(\mathcal{A})).$$

Furthermore, the solution $(\underline{u}, \frac{\partial \underline{u}}{\partial t})$ of (6.3)–(6.7) with initial datum in $\mathcal{D}(\mathcal{A})$ satisfies (6.9). Therefore the energy is decreasing.

2   Asymptotic Behavior

In order to analyze the asymptotic behavior of system (6.3)–(6.7), we shall use Theorems 1.25 and 1.26 which respectively characterize exponential and polynomial stability of a C_0-semigroup of contraction:

Lemma 6.3 (Asymptotic Stability) *The operator $\mathcal{A}$ verifies (1.9) and then the associated semigroup $(S(t))_{t\geq 0}$ is asymptotically stable on $\mathcal{H}$.*

Proof Since $0 \in \rho(\mathcal{A})$ we only need here to prove that $(iw\mathcal{I} - \mathcal{A})$ is a one-to-one correspondence in the energy space $\mathcal{H}$ for all $w \in \mathbb{R}^*$. The proof will be done in two steps: In the first step we will prove the injective property of $(iw\mathcal{I} - \mathcal{A})$ and in the second step we will prove the surjective property of the same operator.

- Suppose that there exists $w \in \mathbb{R}^*$ such that $Ker(iw\mathcal{I} - \mathcal{A}) \neq \{0\}$. So $\lambda = iw$ is an eigenvalue of $\mathcal{A}$, and then let $(\underline{u}, \underline{v})$ be an eigenvector of $\mathcal{D}(\mathcal{A})$ associated to λ. For every j in J we have

$$v_j = \mathbf{i}wu_j, \tag{6.18}$$

$$\partial_x(\partial_x u_j + \beta_j \partial_x v_j) = \mathbf{i}wv_j. \tag{6.19}$$

We have

$$Re\left(\langle \mathcal{A}(\underline{u}, \underline{v}), (\underline{u}, \underline{v})\rangle_{\mathcal{H}}\right) = -\sum_{j\in J}\int_0^{\ell_j} \beta_j \left|\partial_x v_j\right|^2 dx = 0.$$

Then $\beta_j \partial_x v_j = 0$ a.e. on $(0, \ell_j)$.

Let e_j be a K-V edge. According to (6.18) and the fact that $\beta_j \partial_x v_j = 0$ a.e. on $(0, \ell_j)$, we have $\partial_x u_j = 0$ a.e. on Ω_j. Using (6.19), we deduce that $v_j = 0$ on ω_j. Returning back to (6.18), we conclude that $u_j = 0$ on Ω_j.

Putting $y = \partial_x u_j + \beta_j \partial_x v_j = (1 + iw\beta_j)\partial_x u_j$, we have $y \in H^2(0, \ell_j)$ and $\partial_x y = -w^2 u_j$. Hence y satisfies the Cauchy problem

$$\partial_x^2 y + \frac{w^2}{1 + iw\beta_j} y = 0, \quad y(z_0) = 0, \quad \partial_x y(z_0) = 0$$

for some z_0 in Ω_j. Then y is zero on $(0, \ell_j)$ and hence $\partial_x u_j$ and u_j are zero on $(0, \ell_j)$. Moreover u_j and $\partial_x u_j + \beta_j \partial_x v_j$ vanish at 0 and at ℓ_j.

If e_j is a purely elastic edge attached to a K-V edge at one of its ends, denoted by x_j, then $u_j(x_j) = 0$, $\partial_x u_j(x_j) = 0$. Again, by the same way we can deduce that $\partial_x u_j$ and u_j are zero in $L^2(0, \ell_j)$ and at both ends of e_j. We iterate such procedure on every maximal subgraph of purely elastic edges of $\mathcal{G}$ (from leaves to the root), to obtain finally that $(\underline{u}, \underline{v}) = 0$ in $\mathcal{D}(\mathcal{A})$, which is in contradiction with the choice of $(\underline{u}, \underline{v})$.

- Now given $(\underline{f}, \underline{g}) \in \mathcal{H}$, we solve the equation

$$(iw\mathcal{I} - \mathcal{A})(\underline{u}, \underline{v}) = (\underline{f}, \underline{g})$$

or equivalently,

$$\begin{cases} \underline{v} = iw\underline{u} - \underline{f} \\ w^2\underline{u} + \partial_x^2\underline{u} + iw\,\partial_x(\underline{\beta} * \partial_x\underline{u}) = (\partial_x\underline{\beta} * \partial_x\underline{f}) - iw\underline{f} - \underline{g}. \end{cases} \quad (6.20)$$

Let us define the operator

$$A\underline{u} = -\partial_x^2\underline{u} - iw\,\partial_x(\underline{\beta} * \partial_x\underline{u}), \quad \forall \underline{u} \in V.$$

It is easy to show that A is an isomorphism from V onto V' (where V' is the dual space of V obtained by means of the inner product in H). Then the second line of (6.20) can be written as follows:

$$\underline{u} - w^2 A^{-1}\underline{u} = A^{-1}\left(\underline{g} + iw\underline{f} - \partial_x(\underline{\beta} * \partial_x\underline{f})\right). \quad (6.21)$$

If $\underline{u} \in \mathrm{Ker}(\mathcal{I} - w^2 A^{-1})$, then $w^2\underline{u} - A\underline{u} = 0$. It follows that

$$w^2\underline{u} + \partial_x^2\underline{u} + iw\partial_x(\underline{\beta} * \partial_x\underline{u}) = 0. \quad (6.22)$$

Multiplying (6.22) by $\overline{u}$ and integrating over $\mathcal{T}$, then by Green's formula we obtain

$$w^2 \sum_{j \in J} \int_0^{\ell_j} |u_j(x)|^2 \, dx - \sum_{j \in J} \int_0^{\ell_j} |\partial_x u_j(x)|^2 \, dx$$

$$- \mathbf{i} w \sum_{j \in J} \int_0^{\ell_j} \beta_j(x) |\partial_x u_j(x)|^2 \, dx = 0.$$

This shows that

$$\sum_{j \in J} \int_0^{\ell_j} \beta_j(x) |\partial_x u_j|^2 \, dx = 0,$$

which imply that $\underline{\beta} * \partial_x \underline{u} = 0$ in $\mathcal{G}$.

Inserting this last equation into (6.22), we get

$$w^2 \underline{u} + \partial_x^2 \underline{u} = 0, \qquad \text{in } \mathcal{G}.$$

According to the first step, we have that $\mathrm{Ker}(\mathcal{I} - w^2 A^{-1}) = \{0\}$. On the other hand thanks to the compact embeddings $V \hookrightarrow H$ and $H \hookrightarrow V'$ we see that A^{-1} is a compact operator in V. Now thanks to Fredholm's alternative, the operator $(\mathcal{I} - w^2 A^{-1})$ is bijective in V, hence Eq. (6.21) has a unique solution in V, which yields that the operator $(\mathbf{i} w \mathcal{I} - \mathcal{A})$ is surjective in the energy space $\mathcal{H}$. The proof is thus complete.

$\square$

Before stating the main result, we define a property (P) on $\underline{\beta}$ as follows:

$$\forall j \in J, \ \beta'_j, \beta''_j \in L^\infty(0, \ell_j) \quad \text{and} \quad \forall a_k \in \mathcal{V}_{int}, \ \sum_{j \in J_k} d_{kj} \beta'_j(a_k) \le 0. \tag{P}$$

Theorem 6.4 *Suppose that the function $\underline{\beta}$ satisfies property* (P), *then*

(i) *If $\underline{\beta}$ is continuous at every inner node of $\mathcal{T}$, then $(S_d(t))_{t \ge 0}$ is exponentially stable on $\mathcal{H}$.*

(ii) *If $\underline{\beta}$ is not continuous at least at an inner node of $\mathcal{T}$, then $(S_d(t))_{t \ge 0}$ is polynomially stable on $\mathcal{H}$, in particular there exists $C > 0$ such that for all $t > 0$ we have*

$$\left\| e^{\mathcal{A}t} (\underline{u}^0, \underline{u}^1) \right\|_{\mathcal{H}} \le \frac{C}{t^2} \left\| (\underline{u}^0, \underline{u}^1) \right\|_{\mathcal{D}(\mathcal{A})}, \ \forall (\underline{u}^0, \underline{u}^1) \in \mathcal{D}(\mathcal{A}).$$

Proof According to Theorems 1.25, 1.26, and Lemma 6.3, it suffices to prove that for $\gamma = 0$, when $\underline{\beta}$ is continuous at every inner node, or $\gamma = 1/2$, when $\underline{\beta}$ is not continuous at an inner node, there exists $r > 0$ such that

$$\inf_{\|(\underline{u},\underline{v})\|_{\mathcal{H}}, w \in \mathbb{R}} |w|^{\gamma} \|(iw\mathcal{I} - \mathcal{A})(\underline{u}, \underline{v})\|_{\mathcal{H}} \geq r. \tag{6.23}$$

Suppose that (6.23) fails. Then there exists a sequence of real numbers w_n, with $w_n \to \infty$ (without loss of generality, suppose that $w_n > 0$) and a sequence of vectors $(\underline{u}_n, \underline{v}_n)$ in $\mathcal{D}(\mathcal{A})$ with $\|(\underline{u}_n, \underline{v}_n)\|_{\mathcal{H}} = 1$ such that

$$w_n^{\gamma} \|(iw_n\mathcal{I} - \mathcal{A})(\underline{u}_n, \underline{v}_n)\|_{\mathcal{H}} \to 0. \tag{6.24}$$

We shall prove that $\|(\underline{u}_n, \underline{v}_n)\|_{\mathcal{H}} = o(1)$, which contradict the hypotheses on $(\underline{u}_n, \underline{v}_n)$.

Writing (6.24) in terms of its components, we get for every $j \in J$,

$$w_n^{\gamma}(iw_n u_{j,n} - v_{j,n}) =: f_{j,n} = o(1) \quad \text{in } H^1(0, \ell_j), \tag{6.25}$$

$$w_n^{\gamma}(iw_n v_{j,n} - \partial_x(\partial_x u_{j,n} + \beta_j \partial_x v_{j,n})) =: g_{j,n} = o(1) \quad \text{in } L^2(0, \ell_j). \tag{6.26}$$

Note that

$$w_n^{\gamma} \sum_{j \in J} \int_0^{\ell_j} \beta_j(x) |\partial_x v_j(x)|^2 \, dx$$

$$= Re\left(\langle w_n^{\gamma}(iw_n\mathcal{I} - \mathcal{A}_d)(\underline{u}_n, \underline{v}_n), (\underline{u}_n, \underline{v}_n)\rangle_{\mathcal{H}}\right) = o(1).$$

Hence, for every $j \in J$

$$w_n^{\frac{\gamma}{2}} \left\| \beta_j^{\frac{1}{2}} \partial_x v_{j,n} \right\|_{L^2(0,\ell_j)} = o(1). \tag{6.27}$$

Then from (6.25), we get that

$$w_n^{\frac{\gamma}{2}} \left\| \beta_j^{\frac{1}{2}} w_n \partial_x u_{j,n} \right\|_{L^2(0,\ell_j)} = o(1). \tag{6.28}$$

Define $T_{j,n} = (\partial_x u_{j,n} + a_j \partial_x v_{j,n})$ and multiplying (6.26) by $w_n^{-\gamma} q T_{j,n}$ where q is any real function in $H^2(0, \ell_j)$, we get

$$Re \int_0^{\ell_j} iw_n v_{\tilde{a},n} q \overline{T_{j,n}} dx - Re \int_0^{\ell_j} \partial_x T_{j,n} q \overline{T_{j,n}} dx = o(1). \tag{6.29}$$

Using (6.25) we have

$$Re \int_0^{\ell_j} iw_n v_{\tilde{\alpha},n} q \overline{T_{j,n}} dx$$

$$= -Re \int_0^{\ell_j} v_{j,n} q (\partial_x \overline{v_{j,n}} + w_n^{-\gamma} \partial_x \overline{f_{j,n}}) dx + Re \int_0^{\ell_j} iw_n v_{j,n} q \beta_j \partial_x \overline{v_{j,n}} dx$$

$$= -\frac{1}{2} \left[q(x) \left| v_{j,n}(x) \right|^2 \right]_0^{\ell_j} + \frac{1}{2} \int_0^{\ell_j} \partial_x q \left| v_{j,n} \right|^2 dx$$

$$- Im \int_0^{\ell_j} q \beta_j w_n v_{j,n} \partial_x \overline{v_{j,n}} dx + o(1). \tag{6.30}$$

On the other hand, integrating the second term in (6.29) by parts yields

$$Re \int_0^{\ell_j} \partial_x T_{j,n} q \overline{T_{j,n}} dx = \frac{1}{2} \left[q(x) \left| T_{j,n}(x) \right|^2 \right]_0^{\ell_j} - \frac{1}{2} \int_0^{\ell_j} \partial_x q \left| T_{j,n} \right|^2 dx. \tag{6.31}$$

Hence, by substituting (6.30) and (6.31) into (6.29), we obtain

$$\frac{1}{2} \int_0^{\ell_j} \partial_x q \left| v_{j,n} \right|^2 dx + \frac{1}{2} \int_0^{\ell_j} \partial_x q \left| T_{j,n} \right|^2 dx - Im \int_0^{\ell_j} q \beta_j w_n v_{j,n} \partial_x \overline{v_{j,n}} dx$$

$$- \frac{1}{2} \left(\left[q(x) \left| v_{j,n}(x) \right|^2 \right]_0^{\ell_j} + \left[q(x) \left| T_{j,n}(x) \right|^2 \right]_0^{\ell_j} \right) = o(1). \tag{6.32}$$

Lemma 6.5 *The following property holds:*

$$Im \int_0^{\ell_j} q \beta_j w_n v_{j,n} \overline{\partial_x v_{j,n}} dx = o(1). \tag{6.33}$$

Proof Since $w_n^{\frac{\gamma}{2}} \beta_j^{\frac{1}{2}} \partial_x v_{j,n} \to 0$ in $L^2(0, \ell_j)$ and $q \in L^\infty(0, \ell_j)$, it suffices to prove that

$$w_n^{1-\frac{\gamma}{2}} \left\| \beta_j^{\frac{1}{2}} v_{j,n} \right\|_{L^2(0,\ell_j)} = O(1). \tag{6.34}$$

For this, taking the inner product of (6.26) by $iw_n^{1-2\gamma} \beta_j v_{j,n}$ leads to

$$w_n^{2-\gamma} \left\| \beta_j^{\frac{1}{2}} v_{j,n} \right\|_{L^2(0,\ell_j)}^2 = -iw_n^{1-\gamma} \int_0^{\ell_j} \partial_x T_{j,n} \beta_j \overline{v_{j,n}} dx - iw_n^{1-2\gamma} \int_0^{\ell_j} g_{j,n} \beta_j \overline{v_{j,n}} dx. \tag{6.35}$$

Since $\beta_j \in L^\infty(0, \ell_j)$ and $g_{j,n} \to 0$ in $L^2(0, \ell_j)$ we can deduce the inequality

$$- Re(iw_n^{1-2\gamma} \int_0^{\ell_j} g_{j,n}\beta_j \overline{v_{j,n}}dx) \le \frac{1}{4}w_n^{2-\gamma}\left\|\beta_j^{\frac{1}{2}}v_{j,n}\right\|_{L^2(\omega_j)}^2 + o(1). \qquad (6.36)$$

On the other hand, we have

$$- Re\left(iw_n^{1-\gamma} \int_0^{\ell_j} \partial_x T_{j,n}\beta_j \overline{v_{\bar\alpha,n}}dx\right)$$

$$= -Re\left[iw_n^{1-\gamma} T_{j,n}(x)\beta_j(x)\overline{v_{j,n}}(x)\right]_0^{\ell_j}$$

$$+ Re\left[iw_n^{1-\gamma} \int_0^{\ell_j} \left(\beta_j'\partial_x u_{j,n}\overline{v_{j,n}} + \beta_j\beta_j'\partial_x v_{j,n}\overline{v_{j,n}} + \beta_j\partial_x u_{j,n}\partial_x\overline{v_{j,n}}\right)dx\right].$$

$$\qquad (6.37)$$

Using (6.27) and (6.28) we have

$$Re\left(iw_n^{1-\gamma} \int_0^{\ell_j} \beta_j\partial_x u_{j,n}\overline{\partial_x v_{j,n}}dx\right) = o(1). \qquad (6.38)$$

Using again (6.27) and the fact that $\beta_j' \in L^\infty(0, \ell_j)$, we conclude that

$$Re\left(iw_n^{1-\gamma} \int_0^{\ell_j} \beta_j\beta_j'\partial_x v_{j,n}\overline{v_{j,n}}dx\right) \le \frac{1}{4}w_n^{2-\gamma}\left\|\beta_j^{\frac{1}{2}}v_{j,n}\right\|_{L^2(0,\ell_j)}^2 + o(1). \qquad (6.39)$$

Now by (6.26), we obtain after integrating by parts that

$$Re\left[iw_n^{1-\gamma} \int_0^{\ell_j} \beta_j'\partial_x u_{j,n}\overline{v_{j,n}}dx\right]$$

$$= Re\left[w_n^{-\gamma} \int_0^{\ell_{\bar\alpha}} \beta_j'(\partial_x v_{j,n} + w_n^{-\gamma}\partial_x f_{j,n})\overline{v_{j,n}}dx\right]$$

$$= \frac{1}{2}\left[w_n^{-\gamma}\beta_j'(x)|v_{j,n}(x)|^2\right]_0^{\ell_j} - \frac{1}{2}w_n^{-\gamma}\int_0^{\ell_j} \beta_j''|v_{j,n}|^2 dx + o(1).$$

Furthermore, using that $\beta_j'' \in L^\infty(0, \ell_j)$ and that $v_{j,n}$ is bounded, we deduce

$$Re\left[iw_n^{1-\gamma} \int_0^{\ell_j} \beta_j'\partial_x u_{j,n}\overline{v_{j,n}}dx\right] \le \frac{1}{2}\left[w_n^{-\gamma}\beta_j'(x)|v_{j,n}(x)|^2\right]_0^{\ell_j} + O(1).$$

$$\qquad (6.40)$$

Combining (6.38)–(6.40) with (6.37), we get

$$
- Re(iw_n^{1-\gamma} \int_0^{\ell_j} \partial_x T_{j,n} \beta_j \overline{v_{j,n}} dx) \leq -Re\left[iw_n^{1-\gamma} T_{j,n}(x)\beta_j(x)\overline{v_{j,n}}(x)\right]_0^{\ell_j}
$$

$$
+ \frac{1}{2}\left[w_n^{-\gamma}\beta_j'(x)\left|v_{j,n}(x)\right|^2\right]_0^{\ell_j}
$$

$$
+ \frac{1}{4}w_n^{2-\gamma}\left\|\beta_j^{\frac{1}{2}}v_{j,n}\right\|_{L^2(0,\ell_j)}^2 + O(1).
$$

(6.41)

Thus, substituting (6.36) and (6.41) into (6.35) leads to

$$
\frac{1}{2}w_n^{2-\gamma}\left\|\beta_j^{\frac{1}{2}}v_{j,n}\right\|_{L^2(0,\ell_j)}^2 \leq -Re\left[iw_n^{1-\gamma} T_{j,n}(x)\beta_j(x)\overline{v_{j,n}}(x)\right]_0^{\ell_j}
$$

$$
+ \frac{1}{2}\left[w_n^{-\gamma}\beta_j'(x)\left|v_{j,n}(x)\right|^2\right]_0^{\ell_j} + O(1). \quad (6.42)
$$

Summing over $j \in J$,

$$
\sum_{j\in J} w_n^2\left\|\beta_j^{\frac{1}{2}}v_{j,n}\right\|_{L^2(0,\ell_j)}^2 \leq -2\sum_{a_k\in\mathcal{V}_{int}} Re\left(iw_n^{1-\gamma}\overline{v}_n(a_k)\sum_{j\in J_k} d_{kj}\beta_{j_k}(a_k)T_{j_k,n}(a_k)\right)
$$

$$
+ w_n^{-\gamma}\sum_{a_k\in\mathcal{V}_{int}}\left|\overline{v}_n(a_k)\right|^2\sum_{j\in J_k} d_{kj}\beta_{j_k}'(a_k) + O(1).
$$

(6.43)

We have used the continuity condition of $\underline{v}_n$ and the compatibility condition (6.7) at inner nodes and the Dirichlet condition of $\underline{u}$ and $\underline{v}$ at external nodes.

Note that from property (P) we have

$$
\sum_{a_k\in\mathcal{V}_{int}}\left|\overline{v}_n(a_k)\right|^2\sum_{j\in J_k} d_{kj}\beta_j'(a_k) \leq 0. \quad (6.44)
$$

then to conclude, it suffices to estimate

$$
\sum_{a_k\in\mathcal{V}_{int}} Re\left(iw_n^{1-\gamma}\overline{v}_n(a_k)\sum_{j\in J_k} d_{kj}\beta_{j_k}(a_k)T_{j_k,n}(a_k)\right).
$$

Case (i), corresponding to $\gamma = 0$: Here $\underline{\beta}$ is continuous in all inner nodes. It follows that $\sum_{a_k\in\mathcal{V}_{int}} Re\left(iw_n^{1-\gamma}\overline{v}_n(a_k)\sum_{j\in J_k} d_{kj}\beta_{j_k}(a_k)T_{j_k,n}(a_k)\right) = 0$

Then, (6.43) and (6.44) yield

$$w_n^2 \left\| \beta_j^{\frac{1}{2}} v_{j,n} \right\|_{L^2(0,\ell_j)}^2 = O(1)$$

for every $j \in J$, and the proof of Lemma 6.5 is complete for case (i).

Case (ii), corresponding to $\gamma = \frac{1}{2}$: Recall that here the function β is not continuous at some internal nodes. We want estimate the first term in the right hand side of (6.42). To do this it suffices to estimate $Re(iw_n^{1-\gamma} T_{j,n}(x)\beta_j(x_j)\overline{v}_{j,n}(x))$ at an inner node $x = x_j$ when $\beta_j(x_j) \neq 0$. For simplicity and without loss of generality we suppose that x_j is the end of e_j identified to 0 via π_j.

Since β_j is continuous on $[0, \ell_j]$, there exists a positive number $k_j < \ell_j$ such that $\beta_j(x) \neq 0$ on $[0, \ell_j]$. We first prove

$$w_n \left\| v_{j,n} \right\|_{L^2(0,k_j)}^2 = o(1). \tag{6.45}$$

Using Gagliardo–Nirenberg inequality (1.14), (6.27), (6.28) and the boundedness of $v_{j,n}$,

$$\left\| v_{j,n} \right\|_{L^\infty(0,k_j)} \leq C_1 \left\| v_{j,n} \right\|_{L^2(0,k_j)}^{\frac{1}{2}} \left\| \partial_x v_{j,n} \right\|_{L^2(0,k_j)}^{\frac{1}{2}}$$
$$+ C_2 \left\| v_{j,n} \right\|_{L^2(0,k_j)} = O(1),$$

$$w_n^{-\frac{3}{8}} \left\| T_{j,n} \right\|_{L^\infty(0,k_j)} \leq C_1 \left\| w_n^{\frac{1}{4}} T_{j,n} \right\|_{L^2(0,k_j)}^{\frac{1}{2}} \left\| w_n^{-1} T_{j,n}' \right\|_{L^2(0,k_j)}^{\frac{1}{2}}$$

$$+ C_2 w_n^{-\frac{3}{8}} \left\| T_{j,n} \right\|_{L^2(0,k_j)} = o(1).$$

It follows that $iw_n^{-\frac{1}{2}} \left[T_{j,n}(x)\overline{v}_{j,n}(x) \right]_0^{k_j} = o(1)$ and then $w_n^{\frac{1}{2}} \left\| v_{j,n} \right\|_{L^2(0,k_j)}^2 = o(1)$.

Then, we multiply (6.26) by $iw_n^{-\frac{1}{2}} v_{j,n}$ and we repeat exactly the same strategy as before, using (6.26) and $w_n^{\frac{1}{2}} \left\| v_{j,n} \right\|^2 = o(1)$, we obtain (6.45).

We are now ready to estimate $-Re(iw_n^{\frac{1}{2}} T_{j,n}(0)\overline{v}_{j,n}(0))$.

Applying Gagliardo–Nirenberg inequality (1.14) to $w = v_{j,n}$ we obtain, using (6.27),

$$w_n^{\frac{1}{2}} \left\| v_{j,n} \right\|_{L^\infty(0,k_j)} \leq C_1 \left\| w_n^{\frac{3}{4}} v_{j,n} \right\|_{L^2(0,k_j)}^{\frac{1}{2}} \left\| w_n^{\frac{1}{4}} \partial_x v_{j,n} \right\|_{L^2(0,k_j)}^{\frac{1}{2}} + C_2 w_n^{\frac{1}{2}} \left\| v_{j,n} \right\|_{L^2(0,k_j)}$$

$$\leq o(1) + \left\| w_n^{\frac{3}{4}} v_{j,n} \right\| o(1).$$

Using again the Gagliardo–Nirenberg inequality (1.14) with $w = T_{j,n}$,

$$\left\| T_{j,n} \right\|_{L^\infty(0,k_j)} \le C_1 \left\| w_n^{\frac{1}{4}} T_{j,n} \right\|_{L^2(0,k_j)}^{\frac{1}{2}} \left\| w_n^{-\frac{1}{4}} \partial_x T_{j,n} \right\|^{\frac{1}{2}} + C_2 \left\| T_{j,n} \right\|_{L^2(0,k_j)}$$

$$\le o(1) \left\| w_n^{\frac{1}{4}} \partial_x T_{j,n} \right\|_{L^2(0,k_j)}^{\frac{1}{2}} + o(1)$$

$$\le o(1) + o(1) \left\| w_n^{\frac{3}{4}} v_{j,n} \right\|_{L^2(0,k_j)}.$$

Here, we have used (6.26)–(6.28). Then

$$|Re(\mathrm{i} w_n^{\frac{1}{2}} T_{j,n}(0)\overline{v_{j,n}}(0))| \le w_n^{\frac{1}{2}} \left\| v_{j,n} \right\|_{L^\infty(0,\ell_j)} \left\| T_{j,n} \right\|_{L^\infty(0,\ell_j)}$$

$$\le \frac{1}{4} w_n^{\frac{3}{2}} \left\| v_{j,n} \right\|_{L^2(0,k_j)}^2 + o(1) \tag{6.46}$$

and

$$- Re\left[\mathrm{i} w_n^{\frac{1}{2}} T_{j,n}(x)\overline{v_{j,n}}(x) \right]_0^{k_j} \le 2 w_n^{\frac{1}{2}} \left\| v_{j,n} \right\|_{L^\infty(0,\ell_j)} \left\| T_{j,n} \right\|_{L^\infty(0,\ell_j)}$$

$$\le \frac{1}{2} w_n^{\frac{3}{2}} \left\| v_{j,n} \right\|_{L^2(0,k_j)}^2 + o(1). \tag{6.47}$$

Multiplying (6.26) by $\mathrm{i} v_{j,n}$ in $L^2(0, k_j)$ and integrating by parts, we obtain

$$w_n^{\frac{3}{2}} \left\| v_{j,n} \right\|_{L^2(0,k_j)}^2 = -\mathrm{i} w_n^{\frac{1}{2}} \left[T_{j,n}(x)\overline{v_{j,n}}(x) \right]_0^{k_j} + \mathrm{i} w_n^{\frac{1}{2}} \int_0^{k_j} T_{j,n}\partial_x \overline{v_{j,n}} dx + o(1). \tag{6.48}$$

Using (6.27) and (6.28), the second term on the left hand side of (6.48) converges to zero. We conclude, using (6.47) that

$$w_n^{\frac{3}{2}} \left\| v_{j,n} \right\|_{L^2(0,k_j)}^2 = o(1).$$

Return back to (6.46) which yields

$$-Re(\mathrm{i} w_n^{\frac{1}{2}} T_{j,n}(0)\overline{v_{j,n}}(0)) = o(1).$$

We obtain the same result if we suppose that x_j is the end of $e_{j,n}$ identified to ℓ_j via π_j, that is

$$-Re(iw_n^{\frac{1}{2}} T_{j,n}(\ell_j)\overline{v_{j,n}}(\ell_j)) = o(1),$$

and we then conclude that the first term on the right hand side of (6.43) converges to zero.

Then, again, using (6.44), we obtain that

$$\sum_{j \in J} w_n^{\frac{1}{2}} \left\| \beta_j^{\frac{1}{2}} w_n v_{j,n} \right\|_{L^2(0,\ell_j)}^2 = O(1),$$

then

$$w_n^{\frac{3}{2}} \left\| \beta_j^{\frac{1}{2}} v_{j,n} \right\|_{L^2(0,\ell_j)}^2 = O(1)$$

for every $j \in I$, and the proof of Lemma 6.5 is complete for case (ii).   □

Return back to the proof of Theorem 6.4. Substituting (6.33) in (6.32) leads to

$$\frac{1}{2} \int_0^{\ell_j} \partial_x q \, |v_{j,n}|^2 \, dx + \frac{1}{2} \int_0^{\ell_j} \partial_x q \, |T_{j,n}|^2 \, dx$$
$$-\frac{1}{2} \left[q(x) \left(|v_{j,n}(x)|^2 + |T_{j,n}(x)|^2 \right) \right]_0^{\ell_j} = o(1) \qquad (6.49)$$

for every $j \in J$.

Let $j \in J$ such that e_j is a K-V string. First, note that from (6.34), we deduce that

$$\left\| \beta_j^{\frac{1}{2}} v_{j,n} \right\|_{L^2(0,\ell_j)}^2 = o(1).$$

Then, we take $q(x) = \int_0^x \beta_j(s)ds$ in (6.49) to obtain

$$\frac{1}{2} \int_0^{\ell_j} \beta_j \, |T_{j,n}|^2 \, dx - \frac{1}{2} \left(\int_0^{\ell_j} \beta_j(s)ds \right) \left(|v_{j,n}(\ell_j)|^2 + |T_{j,n}(\ell_j)|^2 \right) = o(1).$$
$$(6.50)$$

Since $\frac{1}{2} \int_0^{\ell_j} \beta_j \, |T_{j,n}|^2 \, dx = o(1)$ and $\int_0^{\ell_j} \beta_j(s)ds > 0$, then (6.50) implies

$$|T_{j,n}(\ell_j)|^2 + |v_{j,n}(\ell_j)|^2 = o(1). \qquad (6.51)$$

Therefore (6.49) can be rewritten as

$$\frac{1}{2} \int_0^{\ell_j} \partial_x q \, |v_{j,n}|^2 \, dx + \frac{1}{2} \int_0^{\ell_{\tilde{a}}} \partial_x q \, |T_{j,n}|^2 \, dx$$

$$+ \frac{1}{2} \left(q(0) \, |v_{j,n}(0)|^2 + q(0) \, |T_{j,n}(0)|^2 \right) = o(1). \tag{6.52}$$

Taking $q = x + 1$ in (6.52) implies that $\|v_{j,n}\|_{L^2(0,\ell_j)} = o(1)$ and $\|T_{j,n}\|_{L^2(0,\ell_j)} = o(1)$. Moreover, $\|\partial_x u_{j,n}\|_{L^2(0,\ell_j)} = \|T_{j,n} - \beta_j v_{j,n}\|_{L^2(0,\ell_j)} = o(1)$. Also we have

$$v_{j,n}(0) = o(1) \text{ and } T_{j,n}(0) = o(1). \tag{6.53}$$

Finally, notice that (6.51) signifies that

$$v_{j,n}(\ell_j) = o(1) \text{ and } T_{j,n}(\ell_j) = o(1). \tag{6.54}$$

To conclude, it suffices to prove that

$$\|v_{j,n}\|_{L^2(0,\ell_j)} = o(1) \text{ and } \|\partial_x u_{j,n}\|_{L^2(0,\ell_j)} = \|T_{j,n}\|_{L^2(0,\ell_j)} = o(1) \tag{6.55}$$

for every $j \in I$ such that e_j is purely elastic.

To do this, starting by a string e_j attached at one end to only K-V strings. Using continuity condition of $\underline{v}$ and the compatibility condition at inner nodes, implies that $e_{\tilde{a}}$ satisfies (6.53) or (6.54). Moreover, by taking $q = 1$ in (6.49), we conclude that e_j satisfies (6.53) and (6.54). Then using again (6.49) with $q = x$, we deduce that (6.55) is satisfied by e_j. We iterate such procedure on each maximally connected subgraph of purely elastic strings (from leaves to the root).

Thus $\|(\underline{u}_n, \underline{v}_n)\|_{\mathcal{H}} = o(1)$, which contradicts the hypothesis $\|(\underline{u}_n, \underline{v}_n)\|_{\mathcal{H}} = 1$. □

Remark 6.6

(1) If for every $j \in J$, β_j is continuous on $[0, \ell_j]$ and not vanishes in such interval, then we do not need the property (P) in the Theorem 6.4.

   Indeed (P) is used only to estimate

$$-Re \left(i w_n^{1-\gamma} \int_0^{\ell_j} \partial_x T_{j,n} \beta_j \overline{v_{j,n}} \, dx \right)$$

in (6.35), according to $w_n^{1-\frac{\gamma}{2}} \left\| \beta_j^{\frac{1}{2}} v_{j,n} \right\|_{L^2(0,\ell_j)}$.

This is equivalent to estimate

$$-Re\left(iw_n^{1-\gamma}\int_0^{\ell_j}\partial_x T_{j,n}\overline{v_{j,n}}dx\right)$$

according to $w_n^{1-\frac{\gamma}{2}}\left\|v_{j,n}\right\|_{L^2(0,\ell_j)}$:

$$-Re\left(iw_n^{1-\gamma}\int_0^{\ell_j}\partial_x T_{j,n}\overline{v_{j,n}}dx\right)=-Re\left[iw_n^{1-\gamma}\,T_{j,n}\,\overline{v_{j,n}}\right]_0^{\ell_j}$$

$$+Re\left(iw_n^{1-\gamma}\int_0^{\ell_j}T_{j,n}\overline{\partial_x v_{j,n}}dx\right)$$

$$=-Re\left[iw_n^{1-\gamma}\,T_{j,n}(x)\,\overline{v_{j,n}}(x)\right]_0^{\ell_j}+o(1)$$

as in case (ii) (proof of Theorem 6.4) we prove without using (P) that

$$-Re\left[iw_n^{1-\gamma}\,T_{j,n}(x)\,\overline{v_{j,n}}(x)\right]_0^{\ell_j}\leq\frac{w_n^{2-\gamma}}{4}\left\|v_{j,n}\right\|_{L^2(0,\ell_j)}^2+o(1).$$

(2) We find here the particular cases studied in [39, 53, 54, 56, 87]. Note that concerning the result of polynomial stability in [39, 87] the authors proved that the $\frac{1}{t^2}$ decay rate of solution is optimal when the damping coefficient is a characteristic function.

Conclusion

The mean purpose of this book is to study the stability of infinite-dimensional mathematical models of elastic networks with taking into account different forms of dissipation. The considered networks consist of finitely interconnected strings and beams or their combination. Such study is based on the semigroup approach to elaborate different kinds of stability (specially, exponential and polynomial stability) for the considered systems.

When the dissipation is produced by a thermal effect, by coupling the wave (or the Petrowsky equation) with the heat equation on some components, we establish, under some geometric conditions of the network, that the whole system is exponentially stable. Note that the conditions imposed for a general graph of elastic–thermoelastic beams, to elaborate the exponential stability, can be applied to graphs of elastic–thermoelastic strings, to obtain similar results (Comment 1 of Chap. 2). Furthermore, same geometric conditions applied to a graph of strings with the Kelvin–Voigt damping produce exponential or polynomial decay of the associated semigroup, depending on the regularity of the damping coefficient function at inner nodes.

In the case where external feedbacks are applied on the leafs of a tree of strings and beams, we proved that the associated semigroup (in an appropriate Hilbert space) is exponentially stable if there is no beam following a string (from leaves to the root); we proved the lack of exponential stability (on a simple example) when such condition fails , but the system is always polynomially stable. With the same type of feedback, applied on a tree modeling fluid–structure interaction, with the presence of point mass at inner nodes, we proved some results of stability depending on algebraic conditions on the length of some edges. The exponential stability fails in the case of a circuit attached to an edge or a star with two undamped external nodes.

Some Remarks or Future Developments

- In the case of a graph of elastic–thermoelastic of beams (or strings), what happens if the graph contains a maximal subgraph of thermoelastic components, which is

© The Author(s), under exclusive license to Springer Nature Switzerland AG 2022   135
K. Ammari, F. Shel, *Stability of Elastic Multi-Link Structures*, SpringerBriefs in
Mathematics, https://doi.org/10.1007/978-3-030-86351-7

a circuit or a subgraph of purely elastic components with at least two endpoints in $\mathcal{V}_{ext}$? Part of the answer is in [37], which briefly presented in Comment 2 of Chap. 2 where the authors consider a star-shaped network of elastic–thermoelastic rods.

- As in Chap. 4, we can expect the stability of a graph of strings and beams, some of them are damped by thermal effect, in the two cases of Fourier's law and Cattaneo's law. One can extend the study to the nonlinear case as for a single string or a single beam.

- The presence of circles in the networks considered in this book can be expected. What algebraic and geometric conditions on the network that guaranty the exponential or polynomial stability of the corresponding system? PDEs on cyclic graph can be found in some recent works, such as [65] where the authors studied a nonlinear fractional boundary value problem on a particular graph, namely, a circular ring with an attached edge. We also quote paper [16], where the authors studied the exponential stability of a system of transport equations with intermittent damping on a network of N circles intersecting at a single point O. See also [66] where the authors considered a Schrödinger equation on a tadpole graph.

- Most models presented in this book can benefit, in future studies, from numerical simulation to consolidate the results obtained. See, for example, [37] for elastic–thermoelastic star-shaped network and [12] where the authors studied a nodal feedback stabilization of a star-shaped network of beams and a string. We can found, in some recent papers, numerical method to estimate solutions of EDPs on networks, such as [90?].

References

1. A.B. Abdallah, F. Shel, Exponential stability of a general network of 1-*d thermoelastic materials*. Math. Control Relat. Fields **2**, 1–16 (2012)
2. R.A. Adams, *Sobolev Spaces*. Pure and Applied Mathematics, vol. 65 (Academic, New York, 1975)
3. K. Ammari, Asymptotic behavior of some elastic planar networks of Bernoulli-Euler beams. Appl. Anal. **86**, 1529–1548 (2007)
4. K. Ammari, M. Jellouli, Stabilization of star shaped networks of strings. Diff. Integr. Equ. **17**, 1395–1410 (2004)
5. K. Ammari, M. Mehrenberger, Study of the nodal feedback stabilization of a string-beams network. J. Appl. Math. Comput. **36**, 441–458 (2012)
6. K. Ammari, S. Nicaise, Stabilization of a transmission wave/plate equation. J. Diff. Equ. **249**, 707–727 (2010)
7. K. Ammari, S. Nicaise, *Stabilization of Elastic Systems by Collocated Feedback*. Lecture Notes in Mathematics, vol. 2124 (Springer, Cham, 2015)
8. K. Ammari, F. Shel, Stability of a tree-shaped network of strings and beams. Math. Methods Appl. Sci. **41**, 7915–7935 (2018)
9. K. Ammari, A. Henrot, M. Tucsnak, Asymptotic behaviour of the solutions and optimal location of the actuator for the pointwise stabilization of a string. Asymptot. Anal. **28**, 215–240 (2001)
10. K. Ammari, M. Jellouli, M. Khenissi, Stabilization of generic trees of strings. J. Dyn. Control. Syst. **11**, 177–193 (2005)
11. K. Ammari, M. Jellouli, M. Mehrenberger, Feedback stabilization of a coupled string-beam system. Netw. Heterog. Media **4**, 19–34 (2009)
12. K. Ammari, D. Mercier, V. Régnier, J. Valein, Spectral analysis and stabilization of a chain of serially Euler-Bernoulli beams and strings. Commun. Pure Appl. Anal. **11**, 785–807 (2012)
13. W. Arendt, C.J.K. Batty, Tauberian theorems for one-parameter semigroups. Trans. Am. Math. Soc. **306**, 837–852 (1988)
14. A. Borichev, Y. Tomilov, Optimal polynomial decay of functions and operator semigroups. Math. Ann. **347**, 455–478 (2010)
15. H. Brezis, *Analyse fonctionnelle, théorie et applications* (Masson, Paris, 1983)
16. Y. Chitour, G. Mazanti, M. Sigalotti, Persistently damped transport on a network of circles. Trans. Am. Math. Soc. **369**, 3841–3881 (2017)
17. C. Conca, J. Planchard, M. Vanninathan, *Fluids and Periodic Structures*. Research in Applied Mathematics, no. 38 (Wiley, Masson, 1995)

18. R. Dàger, E. Zuazua, *Wave Propagation, Observation and Control in 1-d Flexible Multi-Structures*. Mathématiques & Applications (Springer, Berlin, 2006)
19. E.B. Davies, *One-Parameter Semigroups*. London Mathematical Society Monographs, vol. 15 (Academic, London, 1980)
20. E.B. Davies, *Spectral Theory and Differential Operators*. Cambridge Studies in Adv. Math., vol. 42 (Cambridge University Press, Cambridge, 1995)
21. B. Dekoninck, S. Nicaise, Control of networks of Euler-Bernoulli beams. ESAIM Control Optim. Calc. Var. **4**, 57–81 (1999)
22. B. Dekoninck, S. Nicaise, The eigenvalue problem for networks of beams. Linear Algebra Appl. **314**, 165–189 (2000)
23. K.-J. Engel, R. Nagel, *One-Parameter Semigroups for Linear Evolution Equations*. Graduate Texts in Mathematics, vol. 194 (Springer, New York, 2000)
24. S. Ervedoza, M. Vanninathan, Controllability of a simplified model of fluid-structure interaction. ESAIM Control Optim. Calc. Var. **20**, 547–575 (2014)
25. L.C. Evan, *Partial Differential Equations*. Graduate Studies in Mathematics, vol. 19 (AMS, Providence, 1998)
26. A. Friedman, *Partial Differential Equations* (Holt, Reinhart and Winston, New York, 1969)
27. L. Garibaldi, M. Sidahmed, Matériaux viscoélastiques: atténuation du bruit et des vibrations. Tech. Ing. **1**, N720 (1999)
28. L.M. Gearhart, Spectral theory for contraction semigroups on Hilbert space. Trans. Am. Math. Soc. **236**, 385–394 (1978)
29. R.F. Gibson, P.R. Mantana, S.J. Hwang, Optimal constrained viscoelastic tape lengths for maximising damping in laminated composites. AIAA J. **29**, 1678–1685 (1991)
30. J.A. Goldstein, *Semigroups of Linear Operators and Applications*. Mathematical Monographs (The Clarendon Press, Oxford University Press, New York, 1985)
31. P. Grisvard, *Elliptic Problems in Nonsmooth Domains*. Mono- Graphs and Studies in Mathematics, vol. 24 (Pitman, Boston, 1985)
32. M. Gugat, *Optimal Boundary Control and Boundary Stabilization of Hyperbolic Systems*. SpringerBriefs in Control, Automation and Robotics (Springer, New York, 2015)
33. M. Gugat, M. Sigalotti, Stars of vibrating strings: switching boundary feedback stabilization. Netw. Heterog. Media **5**, 299–314 (2010)
34. Y.N. Guo, G.Q. Xu, Stability and Riesz basis property for general network of strings. Dyn. Control Syst. **15**, 223–245 (2009)
35. Z.J. Han, L. Wang, Riesz basis property and stability of planar networks of controlled strings. Acta Appl. Math. **110**, 511–533 (2010)
36. Z.-J. Han, E. Zuazua, Decay rates for 1-d heat-wave planar networks. Netw. Heterog. Media **11**, 655–692 (2016)
37. Z. Han, E. Zuazua, *Decay Rates for Elastic-Thermoelastic Star-Shaped Networks*. Networks and Heterogeneous Media, vol. 12 (AMS, Providence, 2017), pp. 461–488
38. S.W. Hansen, Exponential energy decay in a linear thermoelastic rod. J. Math. Anal. Appl. **167**, 429–442 (1992)
39. F. Hassine, Stability of elastic transmission systems with a local Kelvin-Voigt damping. Eur. J. Control **23**, 84–93 (2015)
40. E. Hille, *Functional Analysis and Semigroups*. Amer. Math. Soc. Coll. Publ., vol. 31 (Amer. Math. Soc., New York, 1948)
41. E. Hille, R.S. Phillips, *Functional Analysis and Semigroups* (Colloquium Publications, American Math. Soc., Providence, RI, 1957)
42. F.L. Huang, Characteristic condition for exponential stability of linear dynamical systems in Hilbert spaces. Ann. Diff. Equ. **1**, 43–56 (1985)
43. F. Huang, On the mathematical model for linear elastic systems with analytic damping. SIAM J. Control Optim. **26**, 714–724 (1988)
44. C.D. Johnson, Design of passive damping systems. J. Mech. Des. J. Vib. Acoust. **117**, 171–175 (1995). (50th anniversary combined issue)
45. T. Kato, *Perturbation Theory for Linear Operators* (Springer, New York, 1976)

46. V. Komornik, Rapid boundary stabilization of the wave equation. SIAM J. Control Optim. **92**, 197–208 (1991)
47. M. Krstic, A. Smyshlyaev, *Boundary Control of PDEs: A Course on Backstepping Designs* (SIAM, Philadelphia, 2008)
48. J.E. Lagnese, G. Leugering, E.J.P.G. Schmidt, Modelling of dynamic networks of thin thermoelastic beams. Math. Methods Appl. Sci. **16**, 327–358 (1993)
49. J.E. Lagnese, G. Leugering, E.J.P.G. Schmidt, *Modelling, Analysis and Control of Dynamic Elastic Multi-Link Structures*. System & Control: Foundation & and Applications (Birkhauser, Boston, MA, 1994)
50. G. Lebeau, C. Bardos, J. Rauch, Sharp sufficient conditions for the observation, control and stabilization of waves from the boundary. SIAM J. Control Optim. **30**, 1024–1065 (1992)
51. J.-L. Lions, E. Magenes, *Problèmes non-homogenes et applications*, vol. 1 (Dumond, Paris, 1968)
52. J.-L. Lions, E. Magenes, *Non-homogeneous Boundary Value Problems and Applications*, vol. 1 (Springer, New York, 1972)
53. K. Liu, Z. Liu, Exponential decay of energy of the Euler-Bernoulli beam with locally distributed Kelvin-Voigt damping. SIAM J. Control Optim. **36**, 1086–1098 (1998)
54. K. Liu, Z. Liu, Exponential decay of energy of vibrating strings with local viscoelasticity. Z. Angew. Math. Phys. **53**, 265–280 (2002)
55. Z. Liu, B. Rao, Characterization of polynomial decay rate for the solution of linear evolution equation. Z. Angew. Math. Phys. **56**, 630–644 (2005)
56. Z. Liu, Q. Zhang, Eventual differentiability of a string with local Kelvin-Voigt damping. ESAIM Control Optim. Calc. Var. **23**, 443–454 (2017)
57. Z. Liu, S. Zheng, *Semigroups Associated with Dissipative Systems* (Chapman & Hall/CRC, Boca Raton, 1999)
58. Z.H. Liu, B.Z. Guo, Ö. Morgül, *Stability and Stabilization of Infinite Dimensional Systems with Applications* (Springer, London, 1999)
59. Y. Liu, T. Takahashi, M. Tucsnak, Single input of a simplified fluid-structure interaction model. ESAIM Control Optim. Calc. Var. **19**, 20–42 (2013)
60. K. Liu, Z. Liu, Q. Zhang, Stability of a string with local Kelvin-Voigt damping and non-smooth coefficient at interface. SIAM J. Control Optim. **54**, 1859–1871 (2016)
61. E. Lueders L.H. Fatori, J.E.M. Rivera, Transmission problem for hyperbolic thermoelastic systems. J. Therm. Stresses **26**, 739–764 (2003)
62. A. Marzocchi, J.E.M. Rivera, M.G. Naso, Asymptotic behaviour and exponential stability for a transmission problem in thermoelasticity. Math. Meth. Appl. Sci. **25**, 955–980 (2002)
63. V. Mehandiratta, M. Mehra, A difference scheme for the time-fractional diffusion equation on a metric star graph. Appl. Numer. Math. **158**, 152–163 (2020)
64. V. Mehandiratta, M. Mehra, G. Leugering, Existence and uniqueness results for a nonlinear Caputo fractional boundary value problem on a star graph. J. Math. Anal. Appl. **477**, 1243–1264 (2019)
65. V. Mehandiratta, M. Mehra, G. Leugering, Existence results and stability analysis for a nonlinear fractional boundary value problem on a circular ring with an attached edge: a study of fractional calculus on metric graph. Netw. Heterog. Media **16**, 155–185 (2021). AMS
66. F.A. Mehmeti, K. Ammari, S. Nicaise, Dispersive effects for the Schrödinger equation on a tadpole graph. J. Math. Anal. Appl. **448**, 262–280 (2017)
67. M. Mehra, A. Shukla, G. Leugering, An adaptive spectral graph wavelet method for PDEs on networks. Adv. Comput. Math. **47**, 12 (2021)
68. D. Mercier, V. Régnier, Control of a network of Euler-Bernoulli beams. J. Math. Anal. Appl. **342**, 874–894 (2008)
69. D. Mercier, V. Régnier, Spectrum of a network of Euler-Bernoulli beams. J. Math. Anal. Appl. **342**, 174–196 (2008)
70. T. Meurer, *Control of Higher-Dimensional PDEs: Flatness and Backstepping Designs* (Springer, New York, NY, 2013)

71. G. Mophou, G. Leugering, P.S. Fotsing, Optimal control of a fractional Sturm-Liouville problem on a star graph. Optimization **70**, 659–687 (2021)
72. D. Mugnolo, Gaussian estimates for a heat equation on a network. Netw. Heterog. Media **2**, 55–79 (2006)
73. F.J. Narcowich, G. Chen, S.A. Fulling, S. Sun, Exponential decay of energy of evolution equation with locally distributed damping. SIAM J. Appl. Math. **51**, 266–301 (1991)
74. J. Neças, *Les Méthodes Directes en Théorie des Équations Elliptiques* (Masson, Paris, 1967)
75. S. Nicaise, J. Valein, Stabilization of the wave equation on 1-d networks with delay term in the nodal feedback. Netw. Heterog. Media **2**, 425–479 (2007)
76. L. Nirenberg, On elliptic partial differential equations. Ann. Scoula Norm. Sup. Pisa **13**, 115–162 (1959)
77. K. Ono, A stretched string equation with a boundary dissipation. Kyushu J. Math. **48**, 265–281 (1994)
78. J. Oosveen, *Strongly Stabilizable Distributed Parameter Systems* (SIAM, Philadelphia, 2000)
79. A. Pazy, *Semigroup of Linear Operators and Applications to Partial Differential Equations* (Springer, New York, 1983)
80. J. Prüss, On the spectrum of C_0-semigroups.. Trans. Am. Math. Soc. **284**, 847–857 (1984)
81. R. Racke, J.E.M. Rivera, H.F. Sare, Stability for a transmission problem in thermoelasticity with second sound. J. Therm. Stresses **31**, 1170–1189 (2008)
82. M.D. Rao, Recent applications of viscoelastic damping for noise control in automobiles and commercial airplanes. J. Sound Vib. **262**, 457–474 (2003)
83. M. Renardy, On localised Kelvin-Voigt damping. Z. Angew. Math. Mech. **4**, 280–283 (2004)
84. J.E.M. Rivera, H.P. Oquendo, The transmission problem for thermoelastic beams. J. Therm. Stresses **24**, 1137–1158 (2001)
85. W.M. Schmidt, Simultaneous approximation to algebraic numbers by rationals. Acta. Math. **125**, 189–201 (1970)
86. W.M. Schmidt, *Diophantine Approximation*. Lecture Notes in Mathematics, vol. 785 (Springer, Berlin, 1980)
87. M. Sepúlveda, O.V. Villagrán, M. Alves, J.M. Revera, M.Z. Gary, The asymptotic behavior of the linear transmission problem in viscoelasticity. Math. Nachr. **287**, 483–497 (2014)
88. F. Shel, Exponential stability of a network of elastic and thermoelastic materials. Math. Methods Appl. Sci. **36**, 869–879 (2013)
89. F. Shel, Exponential stability of a network of beams. J. Dyn. Control Syst. **21**, 443–460 (2015)
90. A. Shukla, M. Mehra, G. Leugering, A fast adaptive spectral graph wavelet method for the viscous Burgers' equation on a star-shaped connected graph. Math. Methods Appl. Sci. **43**, 7595–7614 (2020)
91. G. Sklyar, A. Zuyev, *Stabilization of Distributed Parameter Systems: Design Methods and Applications* (Springer International Publishing, Cham, 2021)
92. M. Tucsnak, M. Vanninathan, Locally distributed control for a model of fluid-structure interaction. Syst. Control Lett. **58**, 547–552 (2009)
93. M. Tucsnak, G. Weiss, *Observation and Control for Operator Semigroups*. Birkhäuser Advanced Texts: Basler Lehrbücher (Birkhäuser, Basel, 2009)
94. J. Valein, E. Zuazua, Stabilization of the wave equation on 1-d networks. SIAM J. Control Optim. **48**, 2771–2797 (2009)
95. J.L. Vazquez, E. Zuazua, Large time behavior for a simplified 1d model of fluid-solid interaction. Comm. Partial Differ. Equ. **28**, 1705–1738 (2003)
96. J.L. Vazquez, E. Zuazua, Lack of collision in a simplified 1d model for fluid-solid interaction. Math. Models Methods Appl. Sci. **16**, 637–678 (2006)
97. S.M. Venuti, Modeling, analysis and computation of fluid structure interaction models for biological systems. SIAM Undergrad. Res. Online **3**, 1–17 (2010)
98. J. Von Below, A characteristic equation associated to an eigenvalue problem on c^2-networks. Linear Algebra Appl. **71**, 309–325 (1985)
99. M. Walter, *Simulation-Based Model Reduction for Partial Differential Equations on Networks*. Friedrich-Alexander-Universität Erlangen-Nürnberg, FAU Studies Mathematics and Physics Band 15 (FAU University Press, Erlangen, 2018)

100. J.M. Wang, B.Z. Guo, Riesz basis and stabilization for the flexible structure of a symmetric tree-shaped beam network. Math. Methods Appl. Sci. **31**, 289–314 (2008)
101. A. Wyler, Stability of wave equations with dissipative boundary conditions in a bounded domain. Differ. Integral Equ. **7**, 345–366 (1994)
102. K. Yosida, On the differentiability and the representation of one-parameter semigroups of linear operators. J. Math. Soc. Jpn. **1**, 15–21 (1948)
103. K. Yosida, *Functional Analysis*. Grundlehren der math. Wiss., vol. 123 (Springer, Berlin, 1980)
104. Q. Zhang, Exponential stability of an elastic string with local Kelvin-Voigt damping. Z. Angew. Math. Phys. **61**, 1009–1015 (2010)
105. W. Zhang, W. Liu, Existence and Ulam's type stability results for a class of fractional boundary value problems on a star graph. Math. Methods Appl. Sci. **43**, 8568–8594 (2019)
106. X. Zhang, E. Zuazua, Polynomial decay and control of 1-d hyperbolic-parabolic coupled system. J. Differ. Equ. **204**, 380–438 (2004)
107. E. Zuazua, Null control of a 1-d model of mixed hyperbolic-parabolic type, in *Optimal Control and Partial Differential Equations*, ed. by J.L. Vanaldi et al. (IOS Press, Amsterdam, 2001), pp. 198–210
108. A.L. Zuyev, *Partial Stabilization and Control of Distributed Parameter Systems with Elastic Elements* (Springer, Cham, 2015)

Printed in the United States
by Baker & Taylor Publisher Services